```
var count = 0
```

# INVENTIONS THAT CONNECT

# tekniverse

## by teknikio

A platform to code and make a network of things.
Built for collaboration, digital justice, and open-ended exploration.

**tekniverse.teknikio.com**

Available on iOS, macOS, and
Google Chrome Browser

16

36

**ON THE COVER:**
Seasats' new autonomous boat is a professional-grade ASV born from a decade-old DIY build.

**Photo:** Declan Kerwin

Ashley Biltz, Declan Kerwin, Chris Forde, Len Cullum, Lucas VRTech, Jay Grenier

# Make:

> "And I just can't wait until next Halloween 'cause I've got some new ideas that will really make them scream."
> —Jack Skellington, *The Nightmare Before Christmas*

PRESIDENT
**Dale Dougherty**
dale@make.co

VP, PARTNERSHIPS
**Todd Sotkiewicz**
todd@make.co

## EDITORIAL

EXECUTIVE EDITOR
**Mike Senese**
mike@make.co

SENIOR EDITORS
**Keith Hammond**
keith@make.co

**Caleb Kraft**
caleb@make.co

PRODUCTION MANAGER
**Craig Couden**

CONTRIBUTING EDITOR
**William Gurstelle**

CONTRIBUTING WRITERS
S.A. Applin, Guido Burger, Lydia Sloan Cline, Len Cullum, Kelly Egan, Chris Forde, Greg Gilman, Jaime Grenier, Jay Grenier, Helga Hansen, DJ Harrigan, Bob Knetzger, Helen Leigh, Forrest M. Mims III, Marshall Piros, Elke Schick, Lucas VRTech

## MAKE.CO

ENGINEERING MANAGER
**Alicia Williams**

WEB APPLICATION DEVELOPER
**Rio Roth-Barreiro**

## BOOKS

BOOKS EDITOR
**Patrick DiJusto**

## DESIGN

CREATIVE DIRECTOR
**Juliann Brown**

## GLOBAL MAKER FAIRE

MANAGING DIRECTOR, GLOBAL MAKER FAIRE
**Katie D. Kunde**

MAKER RELATIONS
**Siana Alcorn**

GLOBAL LICENSING
**Jennifer Blakeslee**

## MARKETING

DIRECTOR OF MARKETING
**Gillian Mutti**

COMMUNITY MANAGER
**Dan Schneiderman**

## LEARNING LABS

DIRECTOR OF LEARNING
**Nancy Otero**

## OPERATIONS

ADMINISTRATIVE MANAGER
**Cathy Shanahan**

ACCOUNTING MANAGER
**Kelly Marshall**

OPERATIONS MANAGER & MAKER SHED
**Rob Bullington**

## PUBLISHED BY

MAKE COMMUNITY, LLC
**Dale Dougherty**

Copyright © 2021 Make Community, LLC. All rights reserved. Reproduction without permission is prohibited. Printed in the USA by Schumann Printers, Inc.

Comments may be sent to:
editor@makezine.com

Visit us online:
make.co

Follow us:
🐦 @make @makerfaire @makershed
📘 makemagazine
📷 makemagazine
▶ makemagazine
📺 twitch.tv/make
ⓟ makemagazine

Manage your account online, including change of address:
makezine.com/account
For telephone service call 847-559-7395 between the hours of 8am and 4:30pm CST. Fax: 847-564-9453. Email: make@omeda.com

## Make: Community

Support for the publication of *Make:* magazine is made possible in part by the members of Make: Community. Join us at make.co.

## CONTRIBUTORS
*What would be your dream Halloween costume to DIY?*

**DJ Harrigan**
*Sacramento, CA*
*(The Magic GIF-Ball)*
My dream would be a "working," voice-controlled Inspector Gadget costume!

**S.A. Applin**
*San Mateo, CA*
*(Making Competence, Making Confidence)*
The scariest costume I can think of for 2021? HEAT MISER. Full stop. (e.g. "I'm Mister 46.1C — in Canada.").

**Chris Forde**
*Manchester, England*
*(Botanical Engineering)*
The Betelgeuse costume from the 1988 film starring Michael Keaton.

Issue No. 78, Fall 2021. *Make:* (ISSN 1556-2336) is published quarterly by Make Community, LLC, in the months of February, May, Aug, and Nov. Make: Community is located at 150 Todd Road, Suite 100, Santa Rosa, CA 95407. SUBSCRIPTIONS: Send all subscription requests to *Make:*, P.O. Box 566, Lincolnshire, IL 60069 or subscribe online at makezine.com/subscribe or via phone at (866) 289-8847 (U.S. and Canada); all other countries call (818) 487-2037. Subscriptions are available for $34.99 for 1 year (4 issues) in the United States; in Canada: $43.99 USD; all other countries: $49.99 USD. Periodicals Postage Paid at San Francisco, CA, and at additional mailing offices. POSTMASTER: Send address changes to *Make:*, P.O. Box 566, Lincolnshire, IL 60069. Canada Post Publications Mail Agreement Number 41129568.

# The Imperfect Prototype

### by Dale Dougherty, President of Make: Community

The process of making usually involves creating a prototype that looks nothing like what we wanted it to be or what it eventually turns out to be. You could think of it as a metamorphosis — the cocoon looks nothing like the butterfly it will become. It requires some belief that this work-in-progress will fly or float, but often they don't. Too often, stories we are told about the projects that succeed have been trimmed of important details, time-wasting diversions and near disasters.

That's why I enjoy the kind of stories we can publish in *Make:*. **Mike Flanigan** and his team's journey began 11 years ago as a group of teens, spending a summer in a Tiverton, Rhode Island garage, where they imagined that an unmanned vessel could cross the Atlantic Ocean. They built a 12-foot boat with a DIY guidance system, named it *Scout*, and put it in the water. Guess what happened? Scout stopped working after 1,200 miles and was lost. A failure? No, because that's not the whole story.

The project launched them into what became careers in a new industry around autonomous boats. Several of them moved to San Diego and joined different companies who had access to state-of-the-art technology plus capital funding. They were able to develop their skills and understanding. Yet that garage project, and its close camaraderie, remained with them, like a recurring dream. After seven years, they left their jobs and regrouped to build a newly designed boat that can be used by scientists for tracking whales over long distances (page 36). And their story continues — they aim to launch another ocean-crossing attempt in the upcoming months.

**S.A. Applin** gets at the perception of one's own competence in her article on the so-called Imposter Syndrome (page 10). "As a maker you may have ideas about what constitutes being a 'valid' maker, and whether or not you measure up. Everyone has their own version of what competence means and there are no real 'rules,' so if you would like to change how you define competence as a maker, you can create a new definition that is more inclusive and kind to yourself — and others." In other words, you might think that yourself or what you're making has to be perfect but you can define what is good enough.

**Marshall Piros** describes going to Maker Faire as a 9 year old, and seeing all the amazing things that makers create, and wanting to be part of that. "I always left the Faire feeling inspired and ready to make my own wondrous invention," but that didn't happen for him. Years later, perhaps his lack of success on his own made him a good candidate for Make: Learning Labs, which allowed him to acquire some of the skills and confidence he lacked. We were happy to discover that Marshall could make things with words, a writer who made big leaps from his first draft to his near-perfect final draft, making sense of his own experience and those of others in the program in "Majoring in the Minors" (page 20).

Eric Lander, in an address in June for his swearing-in ceremony as the White House Director of the Office of Science and Technology Policy, talked about "repairing the world," a translation of Tikkun Olam. He talked about how science and technology can improve the human condition "when applied with vision and optimism, with wisdom and humility, with rigor and integrity, and with a commitment to engage and serve everyone." He added: "Importantly, we must continue to repair the world even knowing that we will never perfect it. Still, every day, we must keep trying." The world we occupy is also an imperfect prototype. It needs a lot of work and a lot of people capable of being good enough to contribute. ●

Listen to **Make:Cast**, where I ask makers how they became who they are and what they are trying to do now. makezine.com/makecast

Adobe Stock - artant

# MADE ON EARTH

**Backyard builds from around the globe**

Know a project that would be perfect for Made on Earth?
Let us know: editor@makezine.com

# AFROFUTURISTIC BUILDING BRICKS

EKOWNIMAKO.COM

Can black sculptures made out of Lego lead to liberation for Black people around the world? Toronto-based artist **Ekow Nimako** thinks so, and that has been his guiding principle in making stunning pieces of art, like *Kumbisaleh 3020 CE*, an Afrofuturistic re-imagining of a medieval city in the ancient kingdom of Ghana.

"I reached into the past to essentially propel us into the future," Nimako says. The 30-square-foot architectural sculpture, permanently housed at the Aga Khan Museum in Toronto, was designed using over 100,000 Lego elements to inspire progress, and celebrate Blackness without the backdrop of enslavement, colonization, and violence.

"It's about dreaming or imagining a reality, and through that imagining of what could be, things become reality. Every system we experience through the expanse of human civilization, it's been thought up," the 42-year-old artist says of his effort to use Afrofuturism as a method of Black liberation. "When you think about Afrofuturism, we can think about worlds like [Marvel's] Wakanda — an African nation that has not been touched by colonialism or enslavement, and is exceedingly technologically advanced — that kind of imagination of what could be is what allows things to be."

And Nimako has experienced imagination becoming reality, firsthand, since he fondly remembers playing with Lego at the age of 4 and wishing he could do it for the rest of his life. With 16 exhibitions and three public artworks under his belt since graduating Canada's York University with a BFA in 2010, it appears his childhood dream manifested into a career. Nimako's even got a collaboration with the toy company that sparked his creativity in the works, and his Building Black series will expand in the Fall of 2022, when his next epic world-building exhibition, *Journey of 2,000 Ships*, debuts at the Dunlop Gallery in Regina, Saskatchewan. *—Greg Gilman*

Ekow Nimako

# WORLD MUSIC MAKEZINE.COM/GO/RADIOGLOBE

Have you ever looked at a globe and wondered what people in different locations might be listening to on their radios? Radioglobe answers that by tuning into terrestrial stations from any location you point to on its surface. Turn the globe to focus the 3D-printed navigation "yoke" to a geographic position; its rotary encoders then calculate the longitude and latitude of the position, and use them to access the nearest radio station's internet broadcast and play it via a Raspberry Pi.

Creative technologist and prototyper **Jude Pullen** (judepullen.com) designed and built the system for DesignSpark, a hub that "connects engineers and supports rapid design to manufacturing processes," to celebrate its 10-year anniversary. Pullen decided to extend and modernize radio, collaborating with Don Robson, who wrote the software and handled the database connections.

Pullen tells *Make:* that he applies his early experiences working with wood to 3D printing and assembly. Inspired by nature, he says that he incorporates empathy for materials into his workflow, imagining what it is like to be them, and how they fit together in context. Using this approach, Pullen improved a rough-feeling interaction on an initial prototype of Radioglobe's yoke, changing the orientation of the grain of the printed plastic pieces so they would mesh together more smoothly.

For those that want to build their own globe, Pullen has posted the instructions online: instructables.com/RadioGlobe-Spin-to-Search-Over-Web-Radio-2000-Stat.

The London-based mechanical engineer, chemist, product designer, and artist loves kids' maker projects too, and shares his passions about the joy of making and exploring world cultures with his 5-year-old son. You can find their endeavors at instagram.com/boredsmart. —S.A. Applin

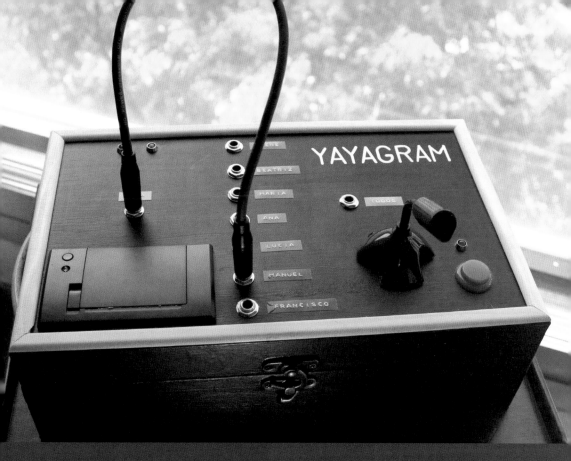

# SWITCHBOARD OPERATOR YAYAGRAM.XYZ

**Manuel "Manu" Lucio Dallo** had a problem — how to keep his family in touch with his 96-year-old grandmother who doesn't own a smartphone. A "computer engineer, maker, and developer at heart," the 35-year-old resident of Burgos, Spain resolved to create something accessible and familiar to his *Yaya* ("grandmother" in Castilian Spanish), that would also work with the Telegram text-messaging service that the grandchildren used.

The resulting "Yayagram" uses tactile, physical buttons, which are easier for Yaya than touch screens and texting. Yayagram converts speech to text that can be received by her grandchildren on their smartphones, and then delivers their text-message responses via a small thermal printer housed inside the device. Yaya can easily access, read, and save them in this paper format. "I tried to adapt the technology to my Yaya and not the other way around," Dallo says.

The machine incorporates an old-fashioned telephone switchboard aspect. To send a message, Yaya inserts an RCA patch cable into a jack for the corresponding grandchild's name (there is also an option for "all"), and then presses the red button while speaking into the microphone. An internal Raspberry Pi captures Yaya's outgoing voice message via the recording software Arecord and ports it to Telegraph wirelessly via their API, a mix of Manu's code, and a Telegram command-line interface library. A single power cable runs the printer and Raspberry Pi inside, keeping things tidy.

"It's certainly a contraption, a box that mixes high tech — Raspberry Pi, Telegram — with low-analog tech — patch cables, connectors, LEDs, a button, and a mic," Dallo says. "This box empowers my Yaya to communicate with all the grandchildren in a very easy way." —*S.A. Applin*

# Making Competence, Making Confidence

Written by S.A. Applin

# WORK YOUR WAY OUT OF IMPOSTER SYNDROME WITH THESE TIPS

**S.A. APPLIN, PH.D.** is an anthropologist whose research explores the domains of human agency, algorithms, AI, and automation in the context of makers, social systems, and sociability. You can find more at @anthropunk and PoSR.org.

Confidence in ourselves and our abilities is something many makers question in themselves from time to time. But a longer lasting dip in self-confidence can keep makers from making and sharing, and that can limit our fun, learning, and enjoyment of sharing our making with others.

*Imposter syndrome* or *imposter phenomenon* and sometimes *tall poppy syndrome* are some ways that society pressures people to limit their self-expression, by stimulating feelings of inhibition and fear about one's abilities. These can stop innovation, as well, if people affected may feel too afraid or anxious of other's opinions of their work to share, market, or develop their projects. A maker with imposter syndrome might feel inhibited to exhibit a project at a local Maker Faire, share a how-to video or Instructable online, or even join a makerspace to learn new skills or to teach others what they know. This limits not just those who wish to make and hold themselves back, but also the rest of us who might benefit from great inventions and contraptions as yet unrealized that could inform new projects.

*Imposter syndrome* is broadly defined as when one does not believe that one is worthy of the position, talent, or opportunity one has naturally or has earned through practice, learning, and developing abilities. The term *imposter phenomenon* was first coined in 1978 by psychologists Dr. Pauline R. Clance and Dr.

Suzanne Imes to describe women's confidence in the workplace, but the definition was described more broadly in 1993 by psychologist Dr. Joe Langford and Dr. Imes as "an experience of feeling incompetent and of having deceived others about one's abilities."

Pretty much anyone trying to do something who feels like they aren't qualified to do it, with or without an educational degree, fancy title, or years of experience, can have a version of imposter syndrome. Makers can develop it as adults, or it can start early in childhood, if a family is unsupportive or dismissive of a child's talents and natural gifts (or loses patience with them for taking apart household things to see how they work). John Gravois, a staff reporter at *The Chronicle of Higher Education*, wrote in 2007 that in the mid-1980s, Dr. Clance and Dr. Gail Matthews conducted a survey on imposter phenomenon, and found that "about 70 percent of people from all walks of life — men and women — have felt like impostors for at least some part of their careers."

## HAMMERED DOWN NAILS

Culturally, imposter syndrome can result from group behavior outside the family, too. Australia, New Zealand, England, and other countries have something called the *tall poppy syndrome*, which evolved in culture to reward those who are self-deprecating, and promote modesty in their achievements so as not to make others feel lesser — or threaten those in power, by being too

# Falling without shame transforms mistakes **into learning**!

capable. People who do not adhere to these social norms can be mocked or otherwise isolated in their social groups. Other cultures have similar notions. In 2017, comparative politics professors Dr. Cornelius Cappelen and Dr. Stefan Dahlburg described cultural ranges of these, such as that in Scandinavia, "the idea that one should never try to be *more*, try to be *different*, or consider oneself more *valuable* than other people is referred to as the Jante mentality," and in Japan there is a popular phrase that "the nail that stands out gets hammered down."

Some research even suggests that the kind of crops a culture grows can create individualistic or group based cultures. In 2014, associate professor of behavioral science Dr. Thomas Talhem and his colleagues conducted a study of farming in China that posited that the needy and finicky tending of rice creates more cooperative group-oriented cultures, while lower-maintenance wheat crops produce more individualistic ones. In these cases, cultural pressures to maintain a status-quo within a group can shape feelings of imposter syndrome as well.

## REDEFINING COMPETENCE

One unfortunate way that imposter syndrome can manifest is when people become afraid that their "stupidity will be discovered." These people make efforts to work extra hard so that people won't know "how stupid they are." Of course, their hard work proves they aren't stupid, but people with imposter syndrome can have a hard time recognizing that they are capable. Some of this has to do with how people define competence. As a maker you may have ideas about what constitutes being a "valid" maker, and whether or not you measure up. Everyone has their own version of what competence means and there are no real "rules," so if you would like to change how you define competence as a maker, you can create a new definition that is more inclusive and kind to yourself — and others.

Dr. Valerie Yates, an expert on imposter syndrome and adult education, created ways to help people understand how they frame their own competence and how to overcome imposter syndrome. The way we perceive our competence impacts our individual manifestations of imposter syndrome, and learning about a framework to understand ways people may define their own competence can help us overcome our own definitions and help us to find new ways to define what is authentic, valuable, and constitutes a "maker." Hint: it's you, just as you are!

Dr. Yates stresses that these are not types of imposter syndrome, but rather **ways that other people have defined competence:**

## TYPES OF COMPETENCE FRAMING

- **The Perfectionist** may not feel worthy if they slip up even a tiny bit in their projects.

- **The Expert** needs to know the answers. Not having "complete" knowledge will stop them from finishing or even beginning a project.

- **The Natural Genius** isn't a "genius," but rather describes people who think they have to be a genius at everything. They are unaware of the learning curve it takes for others to appear competent. People who have this type of competence framing struggle when their first efforts don't measure up to polished versions they see produced by others who are further along the learning curve.

- **The Soloist** only counts competence if they've done a project alone. This type of competence framing disregards group efforts as being meaningful contributions.

- **The Superman/Superwoman** has a competence framing that extends not just from their work or workspace, but into all the roles in their lives, creating impossible goals and enormous pressure.

## WORKING THROUGH IMPOSTER SYNDROME

Dr. Yates says that non-imposters think differently about competence, failure, and fear. She suggests that to overcome feeling like an imposter, people need to "stop thinking like an imposter." That might seem challenging, but here's what she advises:

- *Reframe your thinking. Non-imposters strive to do their best — but they do it because they want to improve, not because they are afraid of being found out.*

💜 **MAKER TIP:** You are doing your best to improve. That's it!

- *Put in the time to learn and see it as growth: non-imposters understand there are times when they have to struggle to understand something or master a new skill.*

💜 **MAKER TIP:** Putting in the time includes setbacks; you will get better!

- *Learn from non-imposters, who know that nothing is going to be perfect the first time — or ever.*

💜 **MAKER TIP:** There is no perfect — enjoy the making!

- *Be OK with falling flat on your face, but the key is to avoid shame about it, and that it's what you do with setbacks, mistakes, and failures that count.*

💜 **MAKER TIP:** Falling without shame transforms mistakes into learning!

- *Remember that it isn't all about you by asking yourself, "What sort of difference would I make if fear was not a factor?" Thinking about how what you make connects with, or can help others, is a great way to appreciate your efforts and see that you are part of a system of cooperation, rather than solving big problems alone.*

💜 **MAKER TIP:** Making is about others, too. Find your connections!

If you'd like to dive in even more, here are other ways you can learn reframing:

- Talk about your fears and confidence issues around making with others. Psychologists Dr. Pauline R. Clance and Dr. Suzanne Imes suggest that finding support in a group of others often leads to a realization that in fact, one is not stupid or inadequate, and that engaging in the compensating behaviors of denying one's abilities, or flattering others to gain acceptance, might be diminishing our true selves. Also, if applicable, an honest eye in a group setting that is supportive can replace the old family group (if it was unsupportive) and give people a new way to perceive their talents and abilities.

- Dr. Yates' lessons on framing competences can help you understand how you categorize your idea of competence. Are you a Soloist? What would it mean for you to learn to expand your idea of competence to include working in a group or on a team project? Even if you are not able to do this yet, being aware of how you frame your competence currently, might help you figure out new ways to frame it that are more supportive and can open you up to accepting competence on your own terms.

- Consider finding a skilled therapist to work through your history and current challenges with making.

- If you have a makerspace, consider running a workshop on how to do things that seem hardest for makers who are fearful. A submission workshop for a local Maker Faire, or an "anything goes share night" or even a "fail night" can create space for people to learn how to more confidently participate on smaller and larger scales.

Hopefully this will help you to reframe competence, realize that making mistakes is part of learning, and that others sometimes struggle, too. You are not alone as a maker — the community is full of support (and we've got your back here at *Make:* and online at makezine.com too). Most importantly, remember why you make (to make!) and enjoy the process of creation, of sharing, and of seeing how what you create can connect to others. ✿

> ## "What sort of difference would I make if **fear was not a factor**?"

## Makerspace Artisan's Asylum: "Embrace failing!"

Massachusetts-based **Artisan's Asylum** has had some history addressing confidence and failure in those coming to its space. Director of Education **Anne Wright** references one memorable instance that provided a "teachable moment" during an outreach program with non-profit Possible Project. The program's high school students, working on their "Build Your Business" prototypes, visited Artisan's Asylum to get feedback and mentorship on their projects from the wealth of diverse making talent in the makerspace. One student had mentioned feeling that "they failed" at something, and Wright flagged it as a worthy discussion point. Being able to teach younger students that failure is normal and an accepted part of the process of design and fabrication was an important lesson for young entrepreneurs. Wright says that Artisan's Asylum was a perfect place for it, being a group that is diverse, supportive, and sees failure as a critical part of the making process. In such an environment, students can take those lessons back to their schools and communities to encourage a broader cultural change, too.

Wright is such a supporter of failure as a part of making that she hopes to have a "Fail Wall" in Artisan's Asylum's new space in the Allston neighborhood of Boston (slated to open in January, 2022). As one of their members, Tim, had said to the students that day, "fail early, fail fast, and fail often," meaning that by doing so, you can move on to the next steps in your project.

# Makerspace Hammerspace:
## "MANY WAYS TO LEARN TOOLS AND SKILLS!"

According to **Dave Dalton**, the "proprietor of **Hammerspace**" (based in Kansas City, Missouri), some makers may have mastery of one set of tools, but are inhibited by others — so much so, that they don't even know where to begin and feel overwhelmed. They might be a whiz in the metal shop, but when confronted with a sewing machine, it may feel insurmountable. Dalton mentions a few paths that people travel to get their projects done when they are stymied by new tools — or more often — new classes of tools.

The first one, which Dalton does not recommend, is to buy the tools you need and to try them alone at home. The path to maker projects often involves failing as an iterative learning method to find out what works. Buying new tools that end up being the wrong tools can be expensive. Someone might be certain that they need a table saw, but end up needing a radial arm saw. That's a pricey mistake! It can hurt emotionally, too, and can feel lonelier (and stupider) on their own trying to figure out what is the right tool.

Dalton says Hammerspace helps overcome that in various ways: by hosting classes; through access to the experts in the makerspace; and through "help me" push-buttons which summon staff members that give advice and training on a tool, or point to the right person who can help. The most critical part of this approach is that the group is there to help someone get through the learning curve.

Ashley Rilty

# the joy of coding

Written by Dale Dougherty

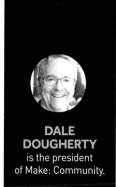

**DALE DOUGHERTY**
is the president
of Make: Community.

## MEET THE TEAM BEHIND THE STREAMING ROBOTICS-EDUCATION PLATFORM, **CodeJoy**

**Kelsey Derringer** and **Matt Chilbert** are co-founders of CodeJoy, producing live, interactive, online shows that give kids live control of their cardboard robots, all from a studio in Pittsburgh, Penn. In doing so, they are creating a new, fresh fusion of education and entertainment through an interactive medium.

**MAKE: Why did you launch CodeJoy?**
**KELSEY**: We wanted to help educators and students push past the financial and emotional

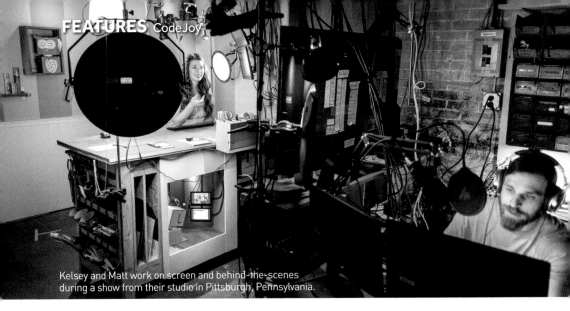

Kelsey and Matt work on screen and behind-the-scenes during a show from their studio in Pittsburgh, Pennsylvania.

barriers of robotics. What if you didn't need to buy expensive robotics equipment to engage in robotics? What if we could provide students an opportunity to do robotics even if their teachers weren't yet comfortable? What if we could show both students and teachers how easy and friendly robotics can be? That is where CodeJoy came from for me — wanting to invite more students and teachers to the robot learning party.

**MATT:** For me, CodeJoy came out of this creative question about the video conference as an entertainment platform. How can we make the video conference look and feel more like a TV show? How could we bring in multiple cameras? How do we bring in pre-recorded video? What does audience engagement look like? How do we hide the seams of this thing to make this truly become a live, interactive TV show?

> WE WANTED TO **HELP EDUCATORS AND STUDENTS** PUSH PAST THE FINANCIAL AND EMOTIONAL BARRIERS OF ROBOTICS.

**MAKE: Why is teaching robotics without a robot helpful?**
**KELSEY:** Robotics is such a great entrance into technology. It's so *real*. It jumps off the screen and grabs you, because something is really happening. I think it's a better entrypoint to coding than straight coding, personally. Plus, robotics breaks stereotypes. When you make a robot that brings you joy, it immediately dispels that idea that coding must be "serious" and coders are only certain types of people. A robot can look like anything! That is so inviting to kids and teachers.

**MATT:** When a robot is a character, well that is automatically interesting.

**KELSEY:** Robots require equipment. But even before that, they require someone to advocate for buying the equipment, and they require teachers who know how to use the equipment. In 2017, in all of the U.S., only something like 36 teachers graduated with CS teaching degrees. Very few teachers feel like they are experts in this field. We don't want to bypass educators, but we do want to support them to see the possibilities here.

**MAKE: What's an example CodeJoy project?**
**MATT:** The one that I'm most excited about right now is our Robot Aerobics show. In this show students learn to code a set of dancing robots. Those robots then teach Kelsey the dance moves created by the students. It's this hilarious feedback loop, honestly. The more they push

the limits, the funnier it is. It makes Kelsey do something ridiculous. It allows them to play first and learn second.

We have to give credit to Mike Cotterman, who's the third member of CodeJoy we brought on officially this year. He really developed CodeJoy.Live, which is our coding platform.

Live-controlled Matt the Robot interacts with students from his "studio under the sink."

**KELSEY:** The platform incorporates two different types of coding built in. There's one that we call batch coding, where kids can create an algorithm and send that sequence to us. And then we can look at the kids' different codes and choose which ones to show and use on the robots to scaffold the learning using their actual code. The other type of coding is live control. With sub-1 second delay, kids can press buttons on their screen and move real robots to do things like drive a robot car or aim and shoot with a cardboard golf putter. So there's less coding knowledge in that, but again, it shows you *why* robotics.

**MAKE: It takes some courage to start something in a pandemic.**
**KELSEY:** In some ways the pandemic has helped. We formed this company with the assumption that we were going to have to explain to people what video conferencing was. And suddenly everybody knew what it was and how to use it, and all the kids had Chromebooks and Wi-Fi, because they needed it for school! And all of this groundwork that we thought we were going to need to build, within a semester it was suddenly *there*.

**MAKE: It's very rewarding to see you bring these experiences to kids who get to feel something, not just learn something, but feel something and have fun.**
**KELSEY:** Well, neither of us actually went to school for computer science or engineering. We both discovered our interests in robotics through fun ourselves — Matt with puppeteering and video production, me as a teacher seeing my students' joy. And that's our goal with CodeJoy — to get kids to have fun with computer science and robotics during our shows, so that they approach their next opportunity to code or make a robot with enthusiasm and confidence. ◕

Students submit code to CodeJoy.Live, controlling the cardboard robots during their live shows.

Kelsey finishes a show with live Q&A from learners.

- Read more from our conversation with Kelsey and Matt at makezine.com/go/codejoy.
- Find CodeJoy online at codejoyeducation.com.

# MINORING IN THE MAJORS

**Written by Marshall Piros**

H ave you ever thought about how makers measure their success? Obviously, making a successful project is a pretty good sign. But, what about the times where a project doesn't work like you envision? I'm talking about when the making process of brainstorming, prototyping, and creating results in, well, *nothing*. A flop. A failure. A pile of parts and pieces spread about. Imagine if makers were assigned "making averages" in the same vein as a baseball player. Batting averages, estimates of a player's success calculated by the number of times they get a hit divided by the number of times they step

## A CRASH COURSE IN MAKING VIA THE MAKE: LEARNING LABS

**MARSHALL PIROS** is a senior in high school who, despite extensively using baseball metaphors in this article, has never actually been to a baseball game.

bat, are easily obtainable statistics that identify truly great swingers. An exceptional batting average is considered around .300, or 3 hits out of every 10 at-bats. Following the same logic, the maker equivalent would be about 3 successful projects out of every 10 attempts. That seems like a strange marker of success, doesn't it? Having only a 30% "success" rate sounds less than ideal. However, the Make: Learning Labs program showed me that, like baseball, numbers like this don't really indicate successes or failures; there is a lot going on behind the scenes of making that a mathematical equation can never capture.

2013 was my first year playing with *Make:*; I was 9 years old and my parents took me to one of the coolest places I had ever been — Maker Faire Bay Area in San Mateo, California. I watched in awe as the movies and toys I loved literally came to life around me; giant replicas of WALL-E and the *Mouse Trap* board game were highlights. I was hooked, this was a true home run. The next year, my dad and I got the equivalent of a maker's Golden Ticket: a free pass to enter the Faire early. We had the entire complex practically to ourselves — a dream come true, which gave me even more time to absorb the magnificence of the DIY creations. I saw miniature Lego dioramas of Gondor from *The Lord of the Rings* and various battles from across the *Star Wars* galaxy. I came across a gigantic steampunk submarine on wheels and befriended one of the life-sized R2-D2s, who followed me around for what seemed like the entire day. Over the next few years, I would re-create *Back to the Future*'s iconic poster (while inside a DeLorean replica), find an action-packed diorama made entirely out of masking tape, watch far too many Mentos shoved into far too many bottles of Diet Coke, and witness the largest bubble blower in the world make a string of soap bubbles about the size of my house. It was all amazing, and I always left the Faire feeling inspired and ready to make my own wondrous invention.

But, that never ended up happening. Despite my love of the finished products at Maker Faire, I was never quite as excited about the process of making, or actually stepping up to the plate.

Up close with R2-D2 at Maker Faire Bay Area 2016, age 12.

I was deathly afraid of a soldering iron for too long. I cannot use scissors to save my life. I never forgot the excitement I felt while at the events, though, which is why I was so interested to learn that *Make:* was sponsoring a program geared towards introducing young adults to making, called the Make: Learning Labs. I made my complete lack of experience very clear in the applications, but I got drafted into the majors anyway. How did that turn out, you might ask? Well, my final project wasn't a 40-foot steampunk fire-breathing octopus or a life-size version of Chutes and Ladders; it's actually this article, highlighting what I've learned both about making and about myself.

The Learning Labs program was an internship designed to introduce a variety of making concepts over a 12-week period, and I joined 12 other interns for the pilot season (a summary of the program can be found at learn.make.co). Truth be told, I was ready to quit after the first week. I felt hopelessly underqualified and struggled to understand the basic concepts from class while the other interns seemed confident enough to start exploring on their own time. Before walking off the field, though, I set up a meeting with Program Director Nancy Otero. To my surprise, she agreed with my assessment that I was a tee-ball player suddenly drafted into the World Series, but she also assured

Marquam Piros, Elizabeth Swartz

me that my skill level didn't really matter that much. When I asked Nancy what she meant, she said that the program was really just about *learning*. We weren't expected to create a world-changing miracle idea, but instead go along with the program and see what we could get out of it. Nancy convinced me that if I was going to be minoring in the majors, I might as well enjoy myself.

With somewhat renewed confidence, I plunged myself back into the first part of the Learning Labs program. This "Inspiration" phase was meant to expose us to a wide array of making concepts on a daily basis for two weeks. We had guest speakers, lectures, homework assignments. The topics varied quickly, from basic coding to constructing paper circuitry. I found that if I stayed focused on learning, as opposed to what my finished product looked like, I could actually enjoy myself. My favorite portions of this phase were the ones that had very clear instructions — beginner makers like me need all the help we can get. Nancy then split us into groups and told us to go create something, and we got to experience the philosophy of making from idea to execution. After this, everyone signed up for three Deep Dives, one- or two-week workshops that went in-depth on a particular making discipline. In other words, speciality training so we could hone our skills for the grand

> ## PRACTICE IS IMPORTANT. NO MATTER HOW SKILLED OR TALENTED A MAKER MIGHT BE, THEY'RE NOT GOING TO BUILD R2-D2 ON THEIR FIRST TRY.

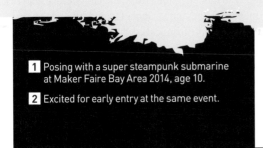

**1** Posing with a super steampunk submarine at Maker Faire Bay Area 2014, age 10.

**2** Excited for early entry at the same event.

finale of the program — final projects. That's where this article came to life: instead of making something, Nancy asked me to write about my experience in the program, and I gladly accepted.

After successfully completing the Make: Learning Labs initiative in one piece (no fingers were lost in the making of this article), here's my post-game analysis:

- **Creativity inspires creativity.** Between the guest lecturers, the mentors, and the other participants, this program showed me that the Make: community is filled to the brim with creative people and wildly interesting ideas. I also found that it doesn't take much for that creativity to rub off.

- **Practice is important.** No matter how skilled or talented a maker might be, they're not going to build R2-D2 on their first try. Becoming proficient in making is no different than learning any other skill; with enough time and practice, you can — and will — improve.

- **There are different positions to play on and off the field.** No team wins with only pitchers or batters. I learned this during my first group project; I may not have been the best maker partner, but I did a great job in making our hard work look presentable. I doubt I'll be at a future Maker Faire as an exhibitor, but I can always see myself in the audience (or even with a press pass).

- **Learning for the sake of learning is actually really useful.** There are no grades, no tests, no GPA to worry about, it's just you and the materials you need. This type of learning is also super flexible about interests, dislikes, and even who can participate. One of the people in the Arduino Deep Dive looked as old as my grandfather. That's really inspiring.

- **Striking out happens way more than hitting a home run.** Whether it's a simple error (making a mistake in one line of code) or something that's a bit more significant (accidentally leaving your heater too close to your Petri dish and melting the agar and growing bacteria), projects usually go through several versions. But even if you only have a 30% success rate, that's still a phenomenal batting average.

And that's the story of how a non-maker took part in an experimental making crash course. Even though I struck out more often than not, I'd still call it a success. I may not become the next breakout inventor who creates exactly what the world needs, but at least I got a new perspective on what those geniuses go through.

# MORE PROJECTS FROM THE INAUGURAL MAKE: LEARNING LABS
learn.make.co

**The Mr. Night Costume**
**Elizabeth Swartz**
Illuminated costume designed to get younger children interested in STEM. Built-in lights activate kids' imaginations as a gateway into astronomy and other subjects.

**SOS WMN**
**Dafne Isadora Cruz Medina,**
**Rodrigo Moreno Oaxaca,** and **Brian Mustafa**
Easy-to-use security device designed to protect women from gender-based crime and violence in public. Sends GPS location and pictures of the situation to emergency contacts when triggered.

**Customizable Pocket AAC**
**Mia Farraday** and **Regina Alatorre Nava**
A low-cost, open source augmentative and alternative communication (AAC) device for autistic individuals.

**Backpack Chair**
**Kate Southern, Gabriela Lopez,** and
**Saul Najera Aguirre**
Portable high chair with massaging function, to help combat chronic pain in adults.

**Corrosion Detector**
**Jannet Galva Acosta** and **Yolotzin Oreday Osorio**
Drone-based service that takes photographs of structures, then analyzes them with machine learning to identify their level of danger.

**Logo Project**
**Ferren Kosciolek**
Learning Labs needed a logo that could capture the spirit of the program, and badges for completing Deep Dives. This project created digital designs for Learning Labs.

See more on the projects from all the participants of the inaugural **Make: Learning Labs** and get more information on how to participate at learn.make.co.

# Fresh Air

Written by Dale Dougherty and Guido Burger

# Making DIY CO₂ monitors to reduce the risk of coronavirus

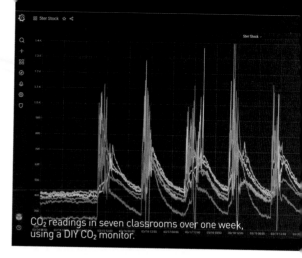

CO₂ readings in seven classrooms over one week, using a DIY CO₂ monitor.

**GUIDO BURGER** is a maker, sailor, and engineer in Stuttgart, Germany. A director at Salesforce, he's also a founder of the nonprofit IoT-Werkstatt, running workshops on sensors, data, AI, and IoT; the designer of the IoT Octopus board; and a contributor to the German *Make:* magazine. @guido_burger

**DALE DOUGHERTY** is publisher of *Make:* Magazine and president of Make: Community. @dalepd

Coronavirus is an airborne disease, like tuberculosis, measles, and the common cold. As of April 30, 2021, the US Centers for Disease Control and Prevention (CDC) accepted the science that *aerosols* were the primary source of transmission. In other words, it's in the air, and not on surfaces, that coronavirus is mostly spread. Tiny droplets or aerosols are exhaled by people and can remain suspended in air for hours and then infect people who have inhaled them. Larger droplets will fall to the ground fairly quickly or be trapped by face masks, but the tiny aerosols (<5μm) are the concern and we can't see them.

Proper indoor ventilation will move fresh air into the room and disperse these aerosols. But how do we know what is proper and safe? That's where *carbon dioxide (CO₂)* sensors come in. In a classroom, for example, filled with students and a teacher, CO₂ will rise if the doors and windows are closed and it will drop if they are left open. CO₂ can be used as a *marker* for proper air ventilation. That familiar feeling that "It's getting stuffy in here" tells you that not only CO₂ is rising, but potentially virus-laden aerosols are rising too.

The questions are when to open the windows and doors, and how long to leave them open? A CO₂ sensing device can provide a warning system that allows you to take action to improve ventilation or reduce the number of people in the room and thus reduce your risk of exposure to Covid-19.

While you can buy commercial CO₂ devices, a DIY CO₂ device is easy to build and you'll gain a better understanding of how such a system works. Plus you can customize it in many ways. Here we'll present a basic instruction set for building a CO₂ device, one in which you can choose which hardware or software to use, as well as choose from a variety of CO₂ sensors. In addition, you can add options to display the readings, light up LEDs to signal a change of status, or sound an alarm.

Recently passed legislation in Nevada and California requires each classroom to have a CO₂ monitor. Wouldn't it be great if students built these monitors for their own schools?

## Introducing Guido Burger, the CO₂ Tech Guru

Our guide for understanding the what, how, and why of CO₂ devices is Guido Burger, a maker and engineer from Germany who last year published the "CO₂ Traffic Light" project in the German edition of *Make:* (Figure Ⓐ on the following page). He's also the developer of the IoT Octopus board.

In 2016, Guido began working with students through a university in Germany. He and professor Klaus-Uwe Gollmer at Trier University

Guido Burger, Adobe Stock-rulizgi

A

of Applied Sciences Umwelt-Campus both had the idea that "We need to get the students in the schools early access to all the technology foundations that we need in this digital age. So how they can work with sensors and the Internet of Things, how they can bring their ideas and innovations to life at an early age."

They happened to decide on $CO_2$ monitoring devices as a project. With Covid-19, the project took on new meaning and Guido began working with students to build $CO_2$ devices and use them in classrooms.

## Understanding $CO_2$

"We are emitting $CO_2$ every time we breathe," said Guido. "Typically, the only sources in a

classroom for $CO_2$ are human beings — while they are talking, speaking, doing their studies. We produce a specific amount of $CO_2$ and the $CO_2$ level rises with the activity. So if we are speaking loud or if we're singing, we are emitting more $CO_2$. At the same time, we're emitting more aerosols as well."

$CO_2$ is a marker for the aerosols in the air but no $CO_2$ device detects the presence of the virus. (That's a project for a device of the future.) "We're making a connection to a potential source of the virus transmission," said Guido. "We're looking at the conditions in which it may be passed on."

$CO_2$ is a gas measured in *parts per million (ppm or p/m)*. In a typical room with nobody in it, $CO_2$

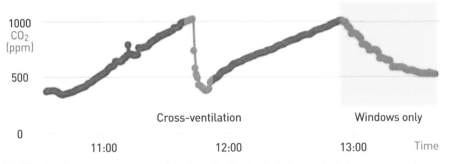

B The $CO_2$ rises in a room over an hour, falls dramatically after both door and windows are opened, and climbs again after they are closed. When only windows are opened, $CO_2$ falls more slowly.

measures around 430ppm, which is the same as outdoor air.

Figure **B** shows readings taken over a 3-hour period in a classroom. *Cross-ventilation* means that a door and window are open and the wind is blowing through; *Windows only* means that only the windows are tilted open a bit.

In Figure B, the $CO_2$ level rises smoothly to about 1000ppm, a level that's widely recognized as a threshold for taking action to improve ventilation. This doesn't mean $CO_2$ levels themselves are harmful, but researchers have calculated, for example, that 1200ppm $CO_2$ means that 2% of the air in the room was inhaled by someone else already (doi.org/10.1034/j.1600-0668.2003.00189.x).

At 1000ppm, we should start opening the door and the window, getting some fresh air in, and when you do that, the $CO_2$ level drops back down fairly quickly. "The wind is blowing into the room and moving the aerosols and the $CO_2$ quickly out of the room," said Guido.

Then the window and doors are closed, perhaps because the room temperature becomes uncomfortable. After about an hour, the $CO_2$ level rises again to previous levels. This time, a window is opened just a little bit. The $CO_2$ level falls, but slowly, because the amount of fresh air coming in through the window is not sufficient. "This could take an hour to get fresh air conditions in the room," said Guido. The $CO_2$ device can give feedback on whether the conditions are changing in the room and how quickly that happens.

A $CO_2$ device can also be used to measure air quality in an air-conditioned room. If the AC is only *recirculating* the air in the room, then $CO_2$ levels will rise. If it is *exchanging* air in the room, then $CO_2$ should remain lower.

## $CO_2$ Sensors

There are several options for $CO_2$ sensors, differing in price and performance. One of the interesting things about $CO_2$ sensors is that they react within seconds to changed conditions. If you breathe anywhere near the sensor, it will report the increase in $CO_2$ almost immediately. Sensors also need to be calibrated, which can be as simple as testing them outdoors first.

# Poorly Ventilated Rooms **Spread Disease**

A 2019 article from researchers in Taiwan looked at the effect of room ventilation on the spread of tuberculosis during an outbreak on a university campus.

**❝Tuberculosis is an airborne disease which spreads through infectious aerosol generated by patients during cough. In an indoor environment, infectious aerosol progressively accumulates and put everyone in the room at risk unless the indoor air is continuously replaced with the fresh outdoor air by ventilation.**

**This study provides the first empirical data showing that improving indoor ventilation to levels with $CO_2$ <1000 ppm is highly effective in controlling a TB outbreak which occurred in poorly ventilated indoor environment.❞**

—Chun-Ru Du, *Effect of ventilation improvement during a tuberculosis outbreak in underventilated university buildings*, ncbi.nlm.nih.gov/pmc/articles/PMC7217216

The researchers found that poorly ventilated rooms at the university were registering levels above 3000ppm but when the $CO_2$ in the room was lowered below 600ppm, the tuberculosis outbreak stopped.

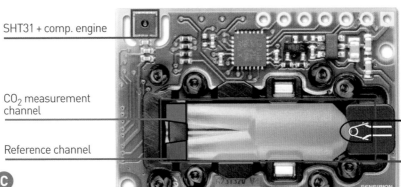

SHT31 + comp. engine

CO₂ measurement channel

Reference channel

Light source

Optical cavity

The Sensirion SCD30 (left) and SCD40 (center) CO₂ sensors, shown here mounted on Guido's Feather breakout board (right).

• **SENSIRION SCD30** is a *nondispersive infrared (NDIR)* optical sensor that costs about $50. Shown in Figures **C** and **D**, this Swiss-made sensor has two channels or tubes that use infrared light to detect CO₂ in the air. The infrared light that's emitted is absorbed by the CO₂ molecules. "The more CO₂ in the tube, the less infrared light is detected," said Guido. The SCD-30 also measures temperature and humidity. It's available on breakout boards including Seeed's Grove and Adafruit's Feather format.

• **SENSIRION SCD40** is a newer, miniature *photoacoustic* CO₂ sensor. It's not dual channel, so it might show a bit more drift in the long run, but it's more rugged and much smaller than the SCD30. Shown in Figure D (it's the tiny cube), it costs about $37. Digi-Key stocks the SCD40 and a breakout board; Guido also makes a tiny breakout, and a larger Feather breakout that accepts either SCD40 or SCD30.

• **WINSEN MH-Z19C** is a Chinese CO₂ NDIR optical sensor with a single channel (Figure **E**). It costs around $35. The advantage of having two channels is greater accuracy —the two readings can be compared — and reliability. But for a low-cost option the MH-Z19 sensor can be used in place of the SCD-30.

• **SENSEAIR S8 LP** — A Swedish NDIR optical sensor (Figure **F**) with the same single channel and footprint as the MH-Z19C but slightly higher price and quality, and better calibration firmware.

## VOC Sensors

Sensors that detect *volatile organic compounds (VOCs)* come in various types that can detect alcohols, aldehydes, methane, carbon monoxide, and other gases or vapors. A VOC sensor is sometimes called an *eCO2* sensor because it estimates an equivalent calculated $CO_2$ level using an algorithm based on the presence of other gases.

The Bosch BME688 sensor, available from Adafruit on a Stemma-format board, is $20. For our purposes, VOC sensors don't effectively measure $CO_2$ but they're great for calculating the *indoor air quality (IAQ)* index, as VOCs are emitted by paint, carpets, hand sanitizers, etc.

Figure 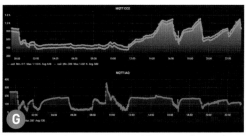 shows a plot of a day in the life of Guido Burger, as measured by a $CO_2$ sensor and a VOC sensor. The orange graph on top follows the $CO_2$ sensor readings while Guido was sitting in his office. Around noontime, he has a phone call and starts talking, and $CO_2$ levels rise to about 1000ppm, and then he opens the window and the value drops. The purple graph shows the VOC sensor, which doesn't change with the rise of $CO_2$ in the orange graph. At times there seems to be a relation between the two, but they're each measuring different things and both can be worth looking at. For instance, the VOC sensor detects changes in air quality while Guido is sleeping even though the $CO_2$ levels in the room remain steady.

## A Warning System

A $CO_2$ device could simply display the ppm reading from the $CO_2$ sensor, but that doesn't provide much context for what the reading means. Guido used three colors of LEDs to create a *$CO_2$ Traffic Light* (*Ampel* in German): "So green light means everything is okay," said Guido.

Comparison of the graphs for a $CO_2$ sensor (orange) and a VOC sensor (purple).

"Yellow means you should open the window. And red light is the ultimate warning."

- **Red** = Above 2000ppm
- **Yellow** = Between 1000 and 2000ppm
- **Green** = Below 1000ppm

Consider changing the threshold values if you wish to be cautious and provide an earlier warning for rising levels. For instance, you might trigger red above 1000ppm, yellow between 800 and 1000ppm, and green below 800ppm.

While individual readings create a simple indicator, we might want to visualize the data over time on a graph, which can show the rate of change. In Figure 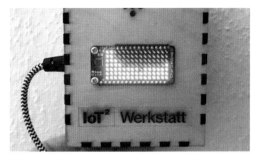, Guido added a matrix display that shows the data as a curve. Now we can see the readings change over time and begin to see how quickly the exchange of fresh air comes in and changes the reading. "This allows us to see ventilation as a dynamic system," explained Guido.

Carter Nelson's *RGB Matrix Portal Room $CO_2$ Monitor* (learn.adafruit.com/matrix-portal-room-co2-monitor) used a large display that anyone in a room could read and understand. The device presents a concise public-health message using

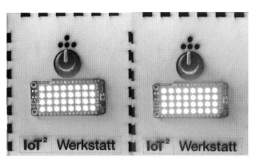

**H** Guido Burger's $CO_2$ Traffic Light, with optional graphical display of $CO_2$ readings over time.

Framed up, Carter Nelson's easy-to-read Room CO₂ Monitor display measures 6½"×11½".

an appropriate emoji with a single word, along with a color and the ppm reading (Figure ①). He used the Adafruit Matrix Portal that provides a plug and play option with an onboard M4-based microcontroller.

## The Elements of a CO₂ Monitor

A CO₂ monitoring device has all the elements of an Internet of Things application: hardware, software, and data. That's a big reason Guido wanted to develop a kit for students to build a device and use it in their schools (Figure ①).

"The kit that the pupils are building these days

is easy to build," he said. "It has a tons of options and it gives us a lot of flexibility." He encourages the kids to create different ways to display the data. "We ask the students to use their creativity in thinking about how they want to present the values, how they want to display this warning, and what visuals or words they want to use," he said.

As Guido began building kits for the students, the Sensirion SCD30 was sold out, in part due to his article on the CO₂ Traffic Light. Guido called the company; they had wondered what caused the unusual spike in demand for individual

## Building a Graph by Hand

Stephan Schulz, a media artist living in Montreal, created a CO₂ device for his daughter, Odessa, to bring to her classroom and monitor ventilation. His simple device used the SCD30 CO₂ sensor connected to an Adafruit Feather M4 Express. His initial version did not store the CO₂ readings but just displayed them. Odessa logged the data manually on a chart, shown here, and at night they'd review the chart and annotate it.

Manual charting of CO₂ readings in a classroom by Stephan and Odessa Schulz.

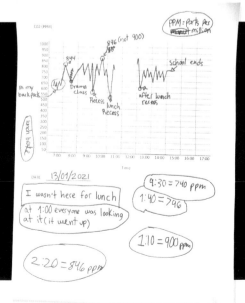

Students building CO$_2$ devices.

The Octopus board with SCD30 Grove sensor.

Octopus with SCD30 Feather sensor.

orders. The sensors are now back in stock but it caused Guido to start designing for different sensors in case he had to find replacements.

## HARDWARE

At the core of a CO$_2$ monitor is a *microcontroller* that connects to the *sensor* and a *display*. The basic microcontroller in our examples is the ESP8266. *Powering* the board and *logging data* are additional considerations; Options 1–3 listed here are powered by USB, Option 4 by battery. The individual components are not expensive but generally add up to $100–$200. You can find the complete parts lists for all these options at makezine.com/go/co2-monitor.

One reason to look at these different options is that some of the components may not be available, which is true not just of the sensors but microcontrollers as well due to chip shortages.

### ■ OPTION 1: IoT Octopus Board

Guido is the developer of the IoT Octopus board and he recommends using it with the Grove SCD30 sensor (Figure **K**). "You can build the device without soldering," he said. "Only two parts are needed because RGB LEDs and ESP8266 are already on the IoT Octopus. And it has the BME680 environment sensor on board as well." Or you can solder the SCD30 Feather sensor instead if you like (Figure **L**).

A separate display is not required because you can use the Octopus onboard LEDs as "traffic lights." If you do want to add a display, consider the Grove 16×2 White on Blue; you can use Grove branch (Y) cables to connect both the sensor and display to the Octopus.

### ■ OPTION 2: DIY Bare Board Kit

This version is the same form-factor as the IoT Octopus and comes as a kit that requires assembly. The red PCB is compatible with several different options for microcontroller, CO$_2$ sensor, and display (Figure **M**). As a kit, it comes with a NodeMCU, Grove connectors, resistors, and

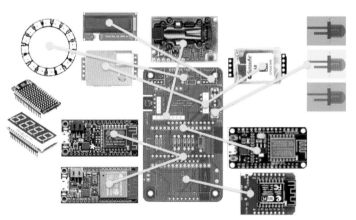

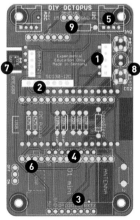

**M** The red DIY Octopus bare board has locations for adding different CO$_2$ sensors (**1**)(**2**), microcontrollers (**3**)(**4**), and displays (**4**)(**5**)(**6**), plus VOC sensor (**7**), discrete LEDs (**8**), and NeoPixel LEDs (**9**).

other components and requires through-hole soldering. After assembly, this handheld device has the SCD30 conveniently stacked on the board, as shown in Figure **N** and on page 24. An optional acrylic case diffuses the green, yellow, and red LEDs so they're visible across the room.

**N** The assembled $CO_2$ device with optional acrylic housing on the left, and with Qwiic 7-segment display on the right.

■ **OPTION 3: Compact and Small $CO_2$ Devices**
You can build a compact $CO_2$ monitor based on an Adafruit Feather microcontroller with the sensor stacked on top, mounted on a FeatherWing board. Guido developed a FeatherWing board with either the two-channel SCD30 (Figure **O**) or the one-channel MHZ-19 $CO_2$ sensors. You can also, as in Option 4, buy equivalent components from Adafruit, Seeed Studios, or SparkFun.

The even smaller SCD40 sensor can also be mounted on a FeatherWing, or piggybacked onto Adafruit's QT Py RP2040 ($10) or QT Py SAMD21 ($8) or the Seeed Xiao SAMD21 ($6). (Note that the RP2040 board can't be programmed using ArduBlock.) A Stemma/Qwiic display or others can be used. You need to do a little soldering of 2×8 header rows but it's simple to build the world's smallest $CO_2$ device (Figure **P**). Guido recently created a fun ruler PCB that also hosts a QT Py and SCD40 (Figure **Q**).

**O** SCD30 sensor mounted on FeatherWing, for stacking on a Feather microcontroller.

Here's a list of possible display options:
• **LED matrix** in different colors (Adafruit Feather CharliePlex 15x7) — for universal output: text, measured values, and trend graph
• **7-segment display** in different colors (Adafruit 7-Segment FeatherWings) — easy-to-read LED digits for measured value output
• **NeoPixel matrix** (Adafruit NeoPixelWing 4x8) —widely visible signal lights (green, yellow, red)
• **NeoPixel ring** (Adafruit NeoPixel 24) and Grove cable (male-jumper to Grove) — to display measurement value (gauge) and traffic light colors simultaneously. You could also use fewer pixels — just increase the parameter delta in the code if necessary (e.g. 200ppm/pixel) and adjust size (e.g. 16 pixels).

■ **OPTION 4: Stephan Schulz's $CO_2$ Monitor**
Stephan Schulz built a $CO_2$ device entirely from Adafruit parts, as a portable battery-powered monitor and data logger that his daughter could take to school (Figure **S**). It's based on a Feather

**P** SCD40 atop an Adafruit QT Py, connected to an Adafruit FeatherWing 128×60 OLED display.

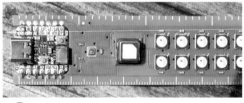

**Q** SCD40 and QT Py on Guido's new ruler PCB.

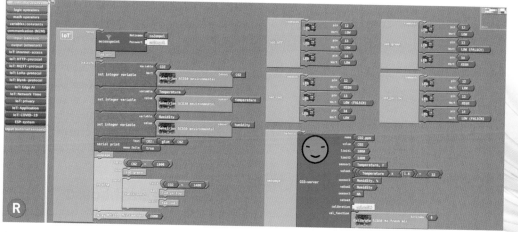

ArduBlock can be used to program the $CO_2$ monitor.

M4 microcontroller, with a FeatherWing OLED display and FeatherWing data logger; he shared his code and bill of materials at github.com/stephanschulz/co2-monitor.

If you'd like to roll your own $CO_2$ device, Carter Nelson wrote an excellent mix-and-match list of Adafruit parts at makezine.com/go/co2-adafruit.

## SOFTWARE

The $CO_2$ monitor can be programmed in Arduino or ArduBlock, a visual block programming editor that is similar to Scratch (Figure **R**). The program consists of a loop that reads the sensor and provides the logic for what to do based on the sensor values, and which LED light to turn on.

The code for the $CO_2$ Traffic Light can be downloaded from github.com/make-IoT/CO2-Ampel in ArduBlock or Arduino. The IoT Werkstatt portable Arduino version includes all needed libraries; download it from the Quick Start guide in English at umwelt-campus.de/en/research/projects/translate-to-englisch-iot-werkstatt.

**The highlights of the code are as follows:**
- Enclose the procedure in a **loop**.
- Create a variable to store for our sensor reading. We called it **CO2**.
- Select the sensor from a library folder: the SCD30 or the MH-Z19.
- Get the reading from a sensor.
  - The sensor is always two blocks. One sends a reading and the other is used for

calibration. Initially, we use the first block just to read a value from the sensor.
- Select from the different values that the sensor provides: $CO_2$, temperature, and humidity.
  - Use the $CO_2$ level only and connect it to the **CO2** variable.
- Create the logic that checks the value and decides what to do, using **if- else** controls and a logic operator:
  - Look for a threshold: if it's less than 1000ppm then turn on green light.
  - If above 1000ppm, turn on the yellow light.
  - If above 1400ppm, turn on the red light.
- At end of loop, add a **2000** millisecond **delay**, which means we are sampling readings every 2 seconds.

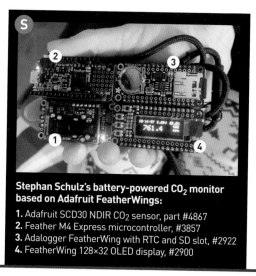

**Stephan Schulz's battery-powered $CO_2$ monitor based on Adafruit FeatherWings:**
1. Adafruit SCD30 NDIR $CO_2$ sensor, part #4867
2. Feather M4 Express microcontroller, #3857
3. Adalogger FeatherWing with RTC and SD slot, #2922
4. FeatherWing 128×32 OLED display, #2900

## The CO₂ Ring

Guido Burger used the tiny SCD40 CO₂ sensor and a NeoPixel ring to create the CO₂ Ring — the world's smallest air-quality wearable detecting CO₂ and VOCs at the same time.

His SCD40 circuit board also features a BME688 VOC sensor and a single NeoPixel. This is stacked on top of a QT Py RP2040 microcontroller, wired to a 12-element NeoPixel ring, and it's all powered by a coin cell battery.

## The CO₂ Canary

Berlin audio journalist and maker Moritz Metz (netzbasteln.de) presented a particularly beautiful version of the CO₂ Traffic Light: He took Guido's project as inspiration for his **CO₂-Narienvogel** (CO₂-nary), which tips off the shelf when the air is bad, like the canary in the coal mine. Code shared at github.com/netzbasteln/co2narienvogel. —*Helga Hansen*

## Installation and Calibration

Install your CO₂ monitor about 1.5m (5') high, away from direct sunlight, windows, and wind gusts. Use a permanent power supply to keep it running 24/7 to see if the room needs active ventilation, and to learn about baseline ventilation, for example during a weekend.

Calibration is automatic. If 420ppm is not shown outdoors, then you can use the calibrations block manually, as auto calibration doesn't work in all cases.

## A Real-Time Dashboard

Guido built a real-time air quality dashboard (Figure **T**) using the open source Grafana data visualization app and MQTT messaging. This requires connecting the device to the internet. Notice there is not one but five traffic lights: CO₂, temperature, humidity, VOCs, and PAX (number of people in the room). "Dry air allows the aerosols to stay even longer in the air," Guido explained.

## Testing Devices in Classrooms

Guido set up seven identical CO₂ devices in seven classrooms for one week in one school. In the chart on page 25, we see CO₂ rising and falling daily from Monday through Friday; school is off Saturday and Sunday. "So you see the values of CO₂ are declining slowly, but as we have seen on the weekend, we are around 400ppm," said Guido. "We have this blue one being an exception." Looks like the blue one still needs to be calibrated.

In Figure **U**, we see the seven classrooms on a Monday, and in Figure **V** we look at just one of them. "Now, the class starts here at 7:30, so the kids are moving in and you see how quickly we are going from a fresh air condition to an alarm situation," Guido explained. "That's literally happening in, let's say 20 minutes. We have the first alarm because the CO₂ reaches 1100ppm. Because this room is equipped with a CO₂ traffic alarm system, the teacher sees a yellow light and opens the windows and door. We see how quickly this value drops down to around 500ppm. Then the teacher closes the window and you see it's going back up again."

These graphs can help a person see how changes in ventilation in the room occur over time, impacted by the actions of people in the

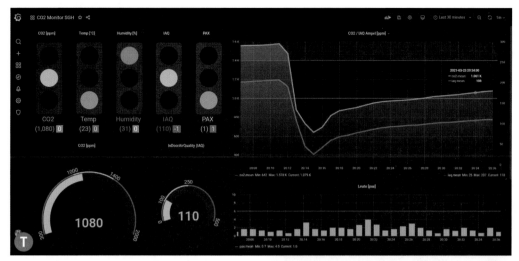

Real-time dashboard using Grafana and MQTT.

room. They can help inform behavior patterns so that teachers, for instance, might understand when to open windows and doors and how long they need to stay open.

That's what makes building a $CO_2$ device such an interesting application. It's not just about gathering data but it's about detecting patterns of $CO_2$ levels indoors, and using messages that help people respond by improving the ventilation — and the health of the people — in the room. ●

Learn more about $CO_2$ monitors and get the parts list to build your own at makezine.com/go/co2-monitor

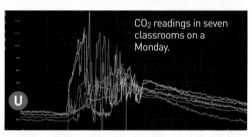

$CO_2$ readings in seven classrooms on a Monday.

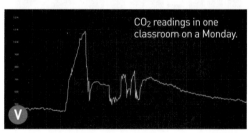

$CO_2$ readings in one classroom on a Monday.

# Is CO₂ Slowing You Down?

Higher $CO_2$ levels are known to impair cognition, another reason for keeping track of it indoors. Because we humans can't detect carbon dioxide ($CO_2$) or its more deadly cousin carbon monoxide ($CO$), we aren't often aware of its effects on us. If you've ever been in a closed conference room after lunch and felt sleepy, it's not digestion that makes you feel that way. It's the rising $CO_2$ level in the room.

While federal guidelines use 5000ppm as the threshold of harm, researchers have found that $CO_2$ levels far below that can affect our thinking. A 2012 study from Lawrence Berkeley Labs found that decision-making suffers at just 1000ppm and dives at 2500ppm (Figure **W**).

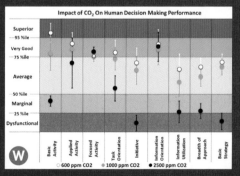

Impact of $CO_2$ on decision-making. Usha Satish, Mark J. Mendell, et al. 2012. doi.org/10.1289/ehp.1104789

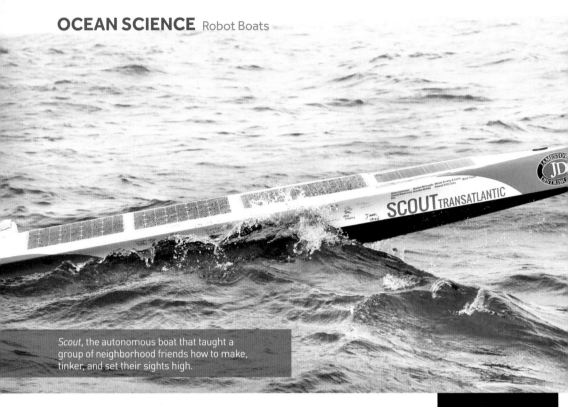

*Scout*, the autonomous boat that taught a group of neighborhood friends how to make, tinker, and set their sights high.

# ACROSS THE SEA, EVENTUALLY

From parents' garages to the high seas, to our new career building ASVs **Written by Mike Flanigan**

**MIKE FLANIGAN** grew up building treehouses, catapults, and other 2×4 contraptions in Tiverton, Rhode Island, where he and his friends launched the autonomous boat *Scout* on its 2013 transatlantic attempt. Mike now lives in San Diego and works with his friends on robots full time, but still loves a good old-fashioned 2×4 project.

Autonomous vehicles are complicated, undoubtedly, but that didn't stop us from endeavoring to be the first to launch a successful autonomous navigation of the Atlantic Ocean. What began as a joke between college kids quickly morphed into a challenge, an obsession, and years later, careers in the industry and a startup of our own.

The first iterations of the boat were designed in 2010 with the goal of traversing the Atlantic autonomously. Since it had never been done before and the journey across the vast ocean was largely unknown and mysterious, we called the project *Scout*. The name was also a shout-out to the *Age of Empires* video game whose scouts we would send off exploring into dark dangerous worlds during the wee hours of the morning.

The fateful joke that started the project was cracked between Dylan Rodriguez and Max Kramers in Max's parents' garage: They'd have to build a robot boat to keep in touch while Max studied in Spain. The project quickly wrapped in more of the neighborhood crew, with Brendan Prior and Dan Flanigan joining the fun. One day I followed Dan, my older brother, into the garage to see what the tinkering was all about.

That garage became the site of a hands-on engineering and maker education for all of us. After our summer jobs we would meet up to sand, solder, and hack away at whatever the current hurdle was. One of my favorite aspects was the frequent pitstops at a whiteboard hanging on the wall, where rapid lessons in basic circuits would come from Dylan, stress and strain from Dan, or lift and hydrodynamics from Max, then after covering the background on why something was being done, we'd jump back to building it. The pace was fast and our target launch date ever approaching. It was full commitment, bordering on over-obsession.

But the project didn't stop with a few kids in a garage. Building an autonomous boat takes a village, or in this case the town of Tiverton, Rhode Island. The project quickly engulfed parents, friends, neighbors, and much of the local community. Some stopped in for an hour of sanding; others made monumental efforts in key technical areas. Tom Schindler knew Dylan from FIRST Robotics and was psyched to add his software skills to the team when we were closing in on launch day. For Tom it was a no-brainer: "These kids were up to something cool, and as soon as I heard about the project, I wanted in." Mike Mills, owner of the local marine and woodworking supply Jamestown Distributors, offered to sponsor the project with all the composite materials we needed.

Along with the moral support and helping hands, there were quite a few chuckles at the endless late nights and the sheer scope of the ambitious endeavor. And with no successful Atlantic crossings to date, despite dozens of attempts by teams on either side of the ocean, ambitious it was.

## Building a Trans-Atlantic ASV

The initial design for *Scout* contained just the basic requisite components: a motor from an R/C airplane, an Arduino backbone, a rudder servo, and a Tupperware to hold it all (see "Your Basic DIY Autonomous Boat," page 41).

But as the project evolved, so did the systems needed for a true autonomous surface vehicle (ASV). While neighbor Greg "Jonesy" Jones was debugging signal noise issues, our young team was busy implementing current sensing, data telemetry, and a hacked GoPro intended to take a few minutes of video each day. The original 3-foot concept craft sank in a pond after an hour, but from that soggy grave it returned and, through six generations, grew to a craft measuring 12 feet.

One of the early iterations of *Scout*.

Much of the complexity involved communication between various technologies. Some components, like the INA219 current sensor breakouts from Adafruit, EM-406 GPS from SparkFun, and LSM303 compass from Pololu, had sample Arduino code that made testing easy. Others, like the Quake Q9612 Iridium transceiver, were more difficult to integrate. At Worcester Polytechnic Institute, Dylan met Ryan Muller who joined the team and spent weekends writing code in the garage for the boat to communicate with the Iridium module. Nowadays, options like the RockBlock module make adding satellite communications to embedded projects much easier.

A landmark day for our team was *Scout*'s first successful autonomous mission navigation, which was supervised over lunch at a nearby sandwich shop. "That was one of our proudest days," recalls Dylan. "One of the few that didn't end in paddling after a runaway craft, debugging the IMU, or falling asleep at the bench." For the team, it was the first day of true autonomy, and a taste of encouraging reward that helped make up for skipped hangouts and late nights.

The original "two-week timeframe" stretched into three years and *Scout* accumulated 2,500 followers on Facebook. Our local paper *The Sakonnet Times* ran several articles on the project including one titled "A Slow Boat to Spain." And finally, on August 24, 2013, after two failed excursions and countless other tests, *Scout* took to the high seas on its third, record-setting attempt. (*Make:* reporter Andrew Terranova wrote about the launch at makezine.com/2013/08/27/transatlantic-drone-takes-to-the-sea.)

A local legend by this point, *Scout*'s public tracking page amassed 750,000 views during its final 5-week journey. It powered through days and most nights, occasionally falling asleep on cloudy days. The page was updated with the latest data every 20 minutes. Friends and followers jokingly complained that it was hard to do their work while

The 12-footer partially built, with the help of neighbors, friends, and the crew.

Early electronics for *Scout*.

*Scout*'s first launch was a well-attended event. A ceremonial cannon was shot to mark the occasion, and Brendan cooked barbecue as Tom finished up the tracking page. Note the hazy conditions which persisted for a week, leading to *Scout*'s first recovery.

Second attempt, a nighttime launch to take advantage of the fully charged batteries. *Scout* traveled 50 miles before its servo failed and it was recovered once again.

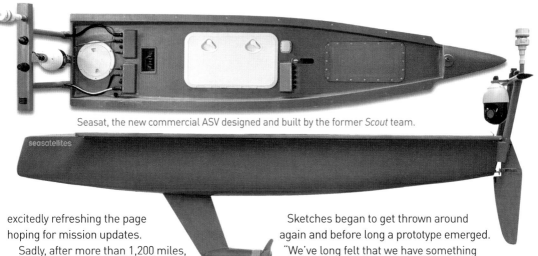

Seasat, the new commercial ASV designed and built by the former *Scout* team.

excitedly refreshing the page hoping for mission updates.

Sadly, after more than 1,200 miles, *Scout* encountered unknown difficulties and fell off the map, so to speak. Our team of excited young engineers, young adults by this point, hung up our hats and parted ways. "We were proud, and devastated, by what we had accomplished," says Dylan, "but we decided it was time to move on to bigger and better things — that's what we thought, at least."

## Going Pro

Dylan got recruited to an autonomous boat company based in San Diego. Max began working for high-performance boat shop Guck Inc., Dan for Reichel-Pugh Yacht Design, Mike for DMC Engineering, and Brendan for Bristol Marine. Tom continued his work with NAVSEA and the team all stayed in touch, one way or another.

But the story didn't end there. Over the next seven years, the notion of returning for another attempt continued to bounce around. And as it did, we were gaining industry skills and experience that would be invaluable for our next endeavor.

As we worked at companies and tracked the progress of the growing ASV scene, we had a common feeling that the costs and complexities of ASV products were too high. Also, it was clear that the companies building these ASVs were ignoring relevant open-source projects that could leapfrog autonomous technology.

Sketches began to get thrown around again and before long a prototype emerged. "We've long felt that we have something to add to the ASV space," says Dan Flanigan. "*Scout* formed the ashes from which Seasats will rise."

Armed with professional experience, this attempt was a bit different. The Seasats project kicked off in 2020 as a side project of Kroova LLC, the engineering consulting company that Dan, Max, and Dylan had launched three years earlier. Kroova gave the team a proper shop and a slim budget with which to work. The cycle of building, testing, and failing resumed, with the occasional success to nudge things onward. By the end of 2020, the team had built a craft that was less expensive and more effective than some of the million-dollar products currently being sold to the US Navy.

That's when we started to understand the opportunity we had. Over the years we had unknowingly accumulated a niche skillset to build just what the market was missing: long-range, autonomous surface vehicles designed for months at sea, but easy enough to launch in minutes, and at a price point that could let the advantages of autonomy reach a wider community.

Today Seasats is in active development. Last winter we ran missions for the nearby Scripps Institute of Oceanography, providing water samples regularly and autonomously. Now we're working with customers in whale tracking, offshore wind power, and the U.S. Navy.

It'd be great to say we came up with an ethos, determined the company vision, and then stuck to that with a laser focus, but the truth is that we started as makers and tinkerers. And honestly today, we're still making and tinkering. But we're no longer in a garage, and we're no longer quite so clueless.

## Trans-Pacific Is Next

Still, the challenge of crossing an ocean lingers in all of our minds. Now located in San Diego we see the Pacific Ocean every day and imagine the challenges and possibilities. Later this year, we will attempt to cross that ocean with one of our Seasats. Let's call it *Scout 2.0*.

"Victory isn't guaranteed, and in many ways the fear of failure is stronger now," says Dylan, "It's easy to be the underdogs, a couple of kids who no one expects to succeed. Now that we look the part, and know enough to be leaders in our field, there's some expectation that we should be able to do this."

But the ocean is a dangerous place with no shortage of obstacles. Stormy seas, overcast days, corrosive salt water, garbage patches, and solar-deck-pooping seagulls all stand in our way. But just like last time, we'll forge forward undaunted. "Once you accept uncertainty as the only certainty," Dylan says, "you begin to realize that it's possible."

Scout team, 2013.

Seasats team, 2021.

Strings to keep pooping birds off the solar deck.

Seasats ASV heading out into the Pacific for offshore testing.

Follow along! Learn more about SeaSats at makezine.com/go/seasats, and track *Scout 2.0*'s upcoming autonomous journey at seasats.com/tracking.

# Your Basic DIY Autonomous Boat

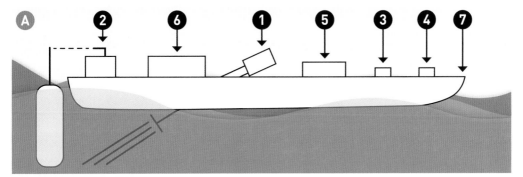

At its most basic form an autonomous boat
is a great DIY project that you can tackle with
basic hobby parts and relatively few components.
The description here outlines a high-level
approach to making a boat that could work well
in bays or ponds.

Why high-level? Because part of the fun of
being a maker is being a believer in guidelines,
not rules! So if you don't have a boogie board,
that's fine, maybe use a couple soda bottles! Or if
you don't want to buy a motor for the propulsion,
but have an old battery powered fan, great, make
it an airboat!

Start with the following core components,
indicated in Figure **A**:

**1 Motor and propeller** Propels the boat
forward. There are lots of options and mounting
configurations, but some of the simplest require
either a long shaft to keep the motor dry while the
propeller is in the water (as shown in Figure A) or
else a waterproof motor.
- **Simple brushed motor with shaft** — Banggood
  #1367490, usa.banggood.com
- **Blue Robotics thruster T200,** bluerobotics.com

**2 Steering servo** Directs the flow of water in
order to steer the boat. There are two common
methods for steering (Figure **B**). *Method 1* uses
a fixed motor and a separate rudder (a piece of
wood or plastic can work well). *Method 2* cuts out
the need for a rudder by simply aiming the output
of the motor; if you have a waterproof motor this
can be a great option. Either configuration could
use a servomotor like this one:
- **Micro continuous rotation servo** Feetech
  FS90R, Pololu 2820, pololu.com

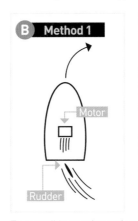
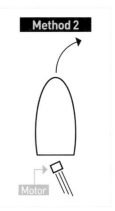

Two possible steering options.

**3 GPS** This is a sensor that triangulates its
position from overhead satellites and can tell
the boat where it is. A great thing about Adafruit's
GPS is that they provide fantastic starter code
and wiring.
- **Ultimate GPS Breakout module** — Adafruit 746,
  adafruit.com

**❹ Compass** This sensor measures Earth's magnetic fields to tell the boat which direction it's pointing. A critical thing to know when trying to navigate at sea! Once again, Adafruit's part has excellent starter code and wiring.

- **Triple-axis accelerometer / magnetometer** Adafruit 1120

> **TIP:** Don't mount the compass anywhere near the motor, motor cables, or big pieces of metal! They can cause interfering magnetic fields that mess up your boat's navigation.

**❺ Controller** A brain of sorts is needed to read the input sensors and, based on commands programmed into its software, direct output signals to the motor and steering servo. Arduinos of all kinds are an excellent option for a low-cost, easy-to-use, controller.

- **Arduino microcontroller boards** arduino.cc

**❻ Battery** You'll want to let your boat go free and wander about in the water, so you'll need a battery to power it. For convenience, choose something with a similar voltage to your controller, motor, and servo to cut down on the number of voltage regulators you'll need.

**❼ Hull** You need your boat to float! This can be a wonderful place for creativity and fun. Use whatever you can get your hands on: a boogie board, a plastic box, soda bottles, your choice.

OK, so with those core components you have all the necessary parts for your very own DIY autonomous boat. Maybe you'll add LEDs (always an awesome decision) or a solar panel to charge your battery. But aside from those fun bits, how do you get your boat to actually start doing things on its own? That's where software comes in.

Figure ⓒ shows a basic loop with the steps needed for an autonomous boat to drive to a waypoint. Seems simple? Well, there are steps left out, but a helpful lesson for software is to keep things in small, testable chunks. If you follow the general steps below and work on making each line work on its own with the necessary sensors and motors then you'll be well on your way to a working boat, minimizing the amount of frustration driven by heaps of code that doesn't work. For software the time-old maker rule is still king: KISS — Keep It Simple, Stupid!

That's one way to get to a functional autonomous boat. Another way we would highly recommend is to look into the fantastic open-source community project ArduPilot and, even better, its ArduBoat section at discuss.ardupilot.org/c/ardurover/arduboat. The community has detailed instructions, components, and software to get you up and ~~running~~ err floating, fast! ⊘

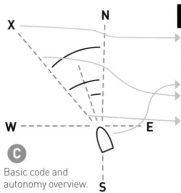

ⓒ Basic code and autonomy overview.

**Pseudo code for autonomous waypoint navigation**

```
waypoint_latitude = 41.726931
waypoint_longitude = 49.948253

Repeat these steps until boat reaches the waypoint
  boat_latitude, boat_longitude = read_gps()
  direction_to_waypoint = calculate_from_boat_and_waypoint_coor
  dinates()
  boat_heading = read_compass()
  steering_angle = calculate_rudder_direction_correction()

  move_rudder(steering_angle)
  turn_motor_on(100%)
```

## Autonomous Boat Do's and Don'ts *Some lessons we learned the hard way.*

**DO:**

- Budget ¼ of your timeline for design, ¼ for build, and ½ for testing. *At least ½.*
- Include a bilge pump — regardless of your confidence in your boat's watertightness.
- Test core components like the motor and rudder servo — for months — on the bench.
- Select important components like solar panels and batteries early in the design phase — before building your boat.
- Compile and use a testing checklist. It's too easy to forget to bring the rudder, the custom-made Allen key, or the spare battery when you head out for testing at 2am.

**DON'T:**

- Set a launch date ASAP. Announce it loudly and confidently.
- Glue all hatches shut just prior to launch, so no one can tamper with your boat.
- Pot all connections in epoxy, test the system, and then rework as needed.
- Find the limit of your telemetry range by driving the boat away from you.
- Keep the keel bulb as light as possible to save on weight.

## Ro, Ro, Roboats *More USV projects we love.* Per ardua ad mare.

### SEA CHARGER

**Damon McMillan**'s solar-powered boat autonomously navigated 2,413 miles from California to Hawaii in 2016 (see "The Little Boat that Could," *Make:* Volume 55, makezine.com/2017/01/25/seacharger). Then he sent it along to New Zealand for the heck of it, and nearly made it; after 6,480 nautical miles the rudder failed, but still! Amazing. Now, Damon says, "I am actively working on a version 2.0 that I hope to be able to sell as a kit for others who want to do this type of thing." seacharger.com

### LOCARB

Inspired by McMillan's feat, **Adrian Li** built a Low-Cost Autonomous Robot Boat for less than $2,000, based on an R/C motor and ESC, a 110W Sunpower solar panel, Arduino Mega microcontroller, PixHawk 1 autopilot, and RockBlock Iridium satellite modem. After two shakedown attempts last year, he launched it May 30 from California to Hawaii and it made 615 miles before succumbing to spring winds and leaks in the battery compartment. Try, try again. Adrian has shared his designs, code, and tips: locarbftw.com

### OPEN OCEAN ROBOTICS

Husband-and-wife explorers **Julie and Colin Angus** spent years doing water expeditions for *National Geographic*, then launched this startup in Tofino, British Columbia, building robust, surf-worthy USVs to gather data on the environment and sea life on Canada's seemingly endless coastlines. openoceanrobotics.com

Damon McMillan, Adrian Li, Open Ocean Robotics

# HOW TO TRACK A SHARK

## Chasing chompers with the latest tech

**Written by Keith Hammond**

**KEITH HAMMOND** is senior editor of *Make:* and spends uncomfortable amounts of time in shark-inhabited waters.

**W**hen scientists tracked great white sharks to a mysterious mid-Pacific hangout, they nicknamed it the White Shark Café — are they eating, or meeting? What's going on out there?

Monterey Bay Aquarium (MBA) researcher Dr. Sal Jorgensen wanted eyes on the scene, so he and engineer Thomas Maughan at sister institution MBARI developed a **Shark Café Camera tracking tag** with funding from the Packard Foundation. Clamped to the dorsal fin, it detects diving and chasing behaviors, then triggers an off-the-shelf Sony Action Cam to record video, and finally pops off and floats for recovery. On its custom PCB you'll find an ATSAMD21 microcontroller (same as Adafruit's Feather M0 boards); a compass/IMU that senses location, acceleration, and even tail beats; an RGB light sensor; and a pressure-transducer depth sensor (Figure Ⓐ). A capacitive sensor detects when it has returned to the surface, then a satellite modem phones home.

If you dare to tag a shark, you can build your

own shark cam: it's open source hardware and software. Maughan (a *Make:* reader, incidentally) taught intern Gabriel Santos a crash course in electronics he dubbed TechFest, then Santos built the camera tags and shared the project at github.com/thommaughan/sharkcafecam. Santos, now a biologist, continues to lead TechFests at CalPoly, teaching the next generation of biologists to use Arduinos and sensors.

Packard's goal was to put better tech in the hands of biologists, so Maughan's team shared their knowledge with Customized Animal Tracking Solutions (cats.is) to help them develop the **CATS Cam** (Figure **B**), now deployed by researchers worldwide on sharks, whales, manta rays, and sea turtles. Jorgensen's work with CATS Cams revealed in 2019 that white sharks in South Africa don't avoid kelp forests — myth busted — but cruise them routinely, bad news for seals and surfers.

Meanwhile, MBA's California tagging program keeps producing amazing discoveries. Turns out even great whites are afraid of something: they'll flee their feeding grounds for up to a year after a single encounter with an orca.

The shark tracker's toolkit also includes **archival tags** (basic data loggers) and **pop-up satellite archival tags (PSATs)** which, like the camera tracker, release from the animal, float, and report their data to the **Argos satellite network** (clsamerica.com/science-with-argos). There are **satellite positioning tags** that report location constantly, **GSM tags** that report to cellular networks, and **acoustic tags** that ping a high-frequency unique ID code that's picked up by receivers listening on the sea floor or afloat.

For do-it-yourselfers, David Mann's Loggerhead Instruments shared their well-regarded **OpenTag** motion and depth logger (loggerhead.com/opentag-motion-datalogger) at github.com/loggerhead-instruments/OpenTag3. And SparkFun now sells an open source **Argos satellite transceiver shield** (sparkfun.com/products/17236) that piggybacks onto their line of

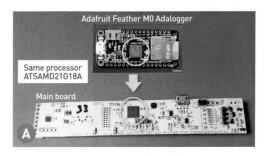

Adafruit Feather M0 Adalogger

Same processor ATSAMD21G18A

Main board

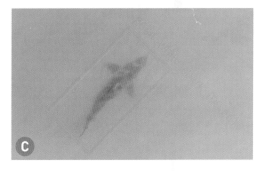

Thing Plus microcontrollers.

In recent years, scientists have increasingly turned to **drone cameras** to spot, follow, and record sharks (researchgate.net/publication/348668975), ranging from typical multirotors like the DJI Phantom to full custom underwater robots like Woods Hole's REMUS AUV. But what's really new is the addition of **artificial intelligence** to detect sharks automatically. The SharkEye project by UC Santa Barbara and San Diego State University (Figure **C**) uses aerial drone video and Salesforce AI to identify great white sharks at a beach with 95% accuracy and send notifications to the local community. ●

# Botanical Engineering

Written and photographed by Chris Forde

Leverage the principles of tensegrity and symmetry to make this seemingly impossible design

What is a planter? Just a container for a plant, and in this instance it's for only a small plant. However, this is not to be any ordinary planter.

This one will make use of symmetry and *tensegrity*. The term, derived from *tension* and *integrity*, is used to describe a structure whose integrity is derived from the balance of tension, not compression. In practice this typically means that structural elements are constrained by cables rather than being stacked atop one another like typical construction methods.

The Tensegrity Planter will be formed from two identical 3D printed elements, either of which can be used to display the plant — but only one at a time — simply by rotation.

This project is fairly easy, assuming you have access to, and are familiar with the use of, a 3D printer. The design can be printed without requiring any supports or rafts. The print time is long, particularly if the elements are printed individually, but this depends upon the bed size of your printer.

## 1. DESIGN

In designing the planter, consideration had to be given to a number of factors. It had to:

- Be capable of supporting and displaying a suitably sized plant
- Be capable of being 3D printed
- Be symmetrical to allow tensegrity
- Have a separate liner to isolate the plant and soil from the structural element, to prevent damage from moisture and minerals
- Have suitable attachment points

The original design started life as a dual stepped pyramid, with additional changes to create the suspension overlap and to extend the first step to create a platform to balance the structure. I used Tinkercad to design the Tensegrity Planter (Figure ) and won First Prize in the Instructables Gardening Challenge.

## 2. PRINT THE PLANTER

Download the 3D print files from youmagine.com/designs/tensegrity-planter. You'll print two copies of *Planter_STL.STL*.

The original model required scaling 400% for my required final size of 161.3×100×96mm.

**TIME REQUIRED:**
A Weekend (Printing + Assembly)

**DIFFICULTY:**
Easy/Intermediate

**COST:**
$20–$75

**MATERIALS**
- » **3D printed parts: planter (2) and liner (1)** Download the free files for printing at youmagine.com/designs/tensegrity-planter. I printed them with PLA filament.
- » **Brass wire, 0.4mm diameter Machine screws, M2.5×10mm (10)**
- » **Two-part epoxy resin, clear**

**TOOLS**
- » **3D printer**
- » **Computer with Cura software** free download at ultimaker.com/software/ultimaker-cura
- » **Sandpaper, various grades**
- » **Wire wool** such as steel wool
- » **Needle file**
- » **Wire cutters**
- » **Level**

**CHRIS FORDE** is an electronics engineer and keen athlete, with multiple Ironman finishes and duathlon/triathlon championships. As a youngster he built a bicycle out of scrap and scavenged parts, and he's been creating ever since, with a variety of tools (hand, FDM, laser, milling, CAD, coding), and materials (wood, plastic, metal, paper, PCB).

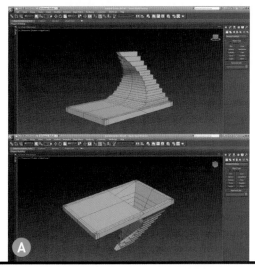

A

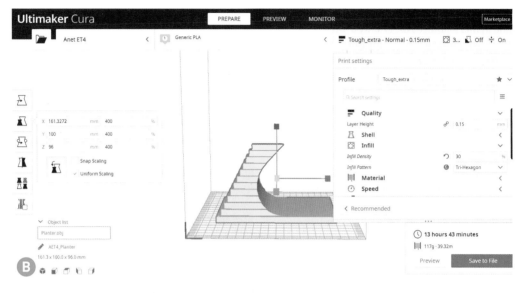

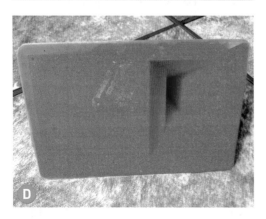

Balancing size and time, I applied the following print settings in Cura (Figure **B**):

- **Layer height:** 0.15mm
- **Infill pattern:** Tri hexagon
- **Infill density:** 30%
- **Base adhesion:** Brim
- **Filament type:** PLA
- **Filament used:** 39.32m
- **Filament weight:** 117g

Print time was 13 hours 43 minutes. As there are two of these elements making up the complete planter (Figures **C** and **D**), that's a total of 27 hours 26 minutes.

## 3. PRINT THE LINER

The other file, *Planter-Insert-STL.STL*, is the liner that the plant will actually sit in, within the planter (Figure **E**).

As a separate element it helps to protect the main structure from the effects of moisture and minerals that will be present in the soil. It also makes it easier to maintain the plant, and the planter itself, by allowing the plant to be removed to prevent interference for adjusting or cleaning.

I printed the liner in black (Figure **F**), so the effects of any discoloration should they occur would be less noticeable, with 100% infill to make it more water and mineral resistant.

Print settings for the liner, also scaled 400%, were as follows:

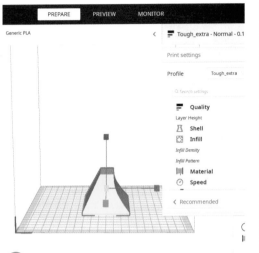

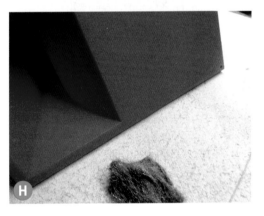

E

- **Layer height:** 0.15mm
- **Infill pattern:** Tri hexagon
- **Infill density:** 100%
- **Base adhesion:** Brim
- **Filament type:** PLA
- **Filament used:** 10.14m
- **Filament weight:** 30g

Print time was 3 hours 25 minutes. Test-fit the liner in the planter (Figure G).

## 4. POST PROCESSING

After printing, remove the brims and smooth any rough edges with sandpaper. Sand the base of the platform, which can also serve as the top, to reduce the print surface striations.

Don't apply too much pressure or use too rough a grade of sandpaper, as you may inadvertently expose the infill pattern or create deep scratches that are difficult to remove. Use progressively finer paper, finishing with wire wool to add a sheen (Figure H). Polishing methods may be applied as required or desired.

For the liner you don't need a polished surface; in fact you want a rough inner surface to improve the epoxy resin adhesion. Use sandpaper to roughen the inside surface of the liner. Then coat it with clear epoxy, to repel water and minerals (Figure I). Apply at least 2 coats, allowing the epoxy to cure between each coat.

Once the epoxy resin has dried, fill the liner

with water and check for any leaks. If any holes are found these can be filled in with epoxy filler prior to use.

## 5. ASSEMBLY

Assembly requires patience and five lengths of wire: 1 short (~11cm) and 4 long (~20cm each).

Start with the suspension points at the apex of the pyramid forms. File a small notch at one point in the circumference of each hole.

Thread the short wire through the hole in the apex, allowing 3cm excess, and fit an M2.5 screw; this will grip the wire and retain the screw in the hole. Wrap the excess wire around the screw at the bottom of the screw head. Allow 5cm of suspension between pyramids, and repeat the same process on the other pyramid.

Now repeat the same process with the 8 holes at the corners using the long wires (Figure **J**), taking care to retain equal tension on each.

Placing a spirit level on the platform will help in balancing the structure (Figure **K**).

## 6. PLANT

Due to the size of the liner, you'll want to populate your planter with a small and slow-growing plant, a succulent or cactus for example.

Put the plant in the liner and fill with soil (Figure **L**). Add a little water if necessary to keep the plant moist. As the liner has no holes, do not overwater the plant as it may damage the liner and/or plant.

Place the liner into the planter.

## 7. DISPLAY

Now that assembly is complete (Figure **M**), it's time to display your work!

## FLOAT SOME IDEAS

There are many different directions you could take this Tensegrity Planter, from simple color changes (Figure **N**), to scaling it up, to adding electronics.

This new version (Figure **O**) is printed in luminous filament and therefore glows in the dark after exposure to light. In addition, 21 NeoPixel LEDs are fitted within the hollow sections and controlled by a micro:bit microcontroller. This allows a number of colors

**TIP:** Tensioning these wires is much easier if a support is used. I created a support out of Lego blocks to support the two main elements with the correct spacing and positioning.

J

K

L

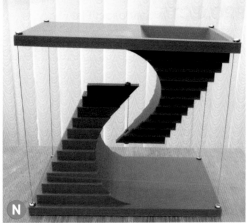

to be programmed, either static or moving (Figures P and Q).

## TIPS AND TRICKS

The trickiest part of the process is the hands-on assembly and ensuring even tension on the wires. I made my tensioning support out of Lego but you could use whatever scrap material you have handy, as long as it supports the two main elements with the correct spacing and positioning, allowing the task of tensioning to be simplified.

My assembled planter stands about 14cm (5.5") tall, 16cm (6.3") long, and 10cm (4") wide, and the liner is just 5.5cm (2") tall, 6.5cm (2.5") long, and 8.8cm (3.5") wide.

If you scale the model up, be aware that the guide holes, wires, and bolts have to be scaled up to support the weight of the plant and soil plus the upper structure. Scaling down will require smaller bolts but it will also make it difficult to accommodate a plant!

One final thing I've just come across: Extremes of temperature can cause the wires to either sag or come loose. Therefore, keep your Tensegrity Planter out of direct sunlight. ◎

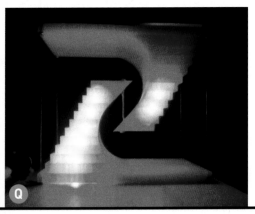

Read more at makeprojects.com/project/tensegrity-planter
Share your build! makeprojects.com

# Grit Stash

Written and photographed by Len Cullum

**TIME REQUIRED:** 1–2 Days

**DIFFICULTY:** Easy

**COST:** $100

## MATERIALS

» **Veneer core plywood, ¾" (18mm) thick, half sheet (4'×4')** You could use regular, 5-ply ¾" plywood but you'll get much better results using a 9- or 11-ply, veneer core product. Look for Baltic birch, Russian birch, or ApplePly.
» **Veneer core plywood, ¼" (6mm) thick, full sheet (4'×8')**
» **Maple board, 1×4 (nominal), 9' long** True dimensions are ¾"×3½".
» **Wood glue**
» **Trim screws, 1⅝" (about a dozen)**
» **Drywall screws, ¾" (about 4 dozen)**
» **Nails, #18×¾"** for the back
» **Wood screws, #6×¾" (2)**
» **Flat washers, #6 (2)**
» **Finish washers, #6 (2)**
» **Hacksaw blade, 12"** for the tearing drawer

## TOOLS

» **Table saw**
» **Dado blades** for table saw
» **Bandsaw (optional)**
» **Miter/chop saw, handsaw, and/or miter box**
» **Hammer**
» **Nailset (optional)**
» **Drill**
» **Forstner bit, ⅝"**
» **Sharp chisel (optional)** A hobby knife will do.
» **Sander**
» **Tape measure**
» **Adjustable square**
» **Waxed paper**
» **Jointer (optional)**
» **Planer (optional)**
» **Drill press (optional)**

**LEN CULLUM** is a woodworker in Seattle, Washington. He also wrote "Japanese Toolbox," "$30 Micro Forge," "Salt and Pepper Well," and more projects and Skill Builders at makezine.com/author/len-cullum.

# Banish the random sheets and mystery grits, and get organized with a sandpaper flat file

**Sandpaper is one of those things that can be a challenge to store.** Bunches of random loose sheets thrown in a box or drawer, rubbing together and losing their grit, or inside cardboard sleeves obfuscating how many sheets you have left, or worse, mysterious partial sheets with no grit numbers on the back.

Having a simple way to store every grit in its own place makes it a lot easier. This sandpaper flat file has enough drawers for all the grits I keep on hand, from 60 up to 400, a couple of drawers for the super high grits from 1000–2000, plus a drawer for sanding sticks and blocks, and one for lightly used pieces that still have a bit of life in them. And lastly a hidden bonus: the narrowest drawer is actually a straightedge tearing board for quickly making half sheets (Figure **A**).

**A**

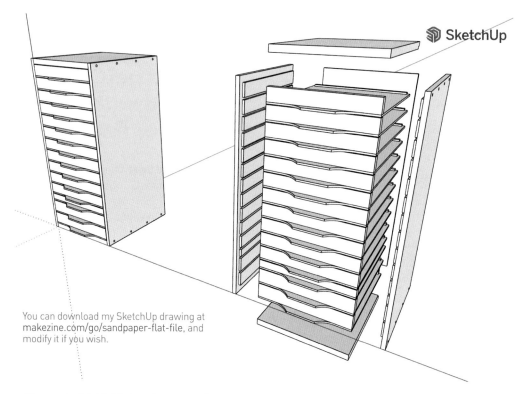

SketchUp

You can download my SketchUp drawing at makezine.com/go/sandpaper-flat-file, and modify it if you wish.

You can build this file in a day or two, for your own workshop or makerspace, or as a gift for anyone who does wood, fiberglass, resin, or other fabrication and finish work.

I use a jointer and planer to dimension the maple wood for the drawer fronts to exactly ⅝" thick. But if you don't have access to these tools, you can work with off-the-shelf ¾" wood instead. Select the straightest piece(s) you can, and rip them to size on the table saw. To accommodate the extra thickness, you'll just cut the carcass, front rabbet, and drawer bottoms a bit deeper:

- Add ⅛" to the carcass pieces, making them 13⅛" deep
- Change the front rabbet from ²¹/₃₂" to ¾" wide
- Make the drawer bottoms ⅛" deeper, from 13½" to 13⅝".

## BUILD YOUR SANDPAPER FLAT FILE

### 1. CUT THE CARCASS PIECES

The first thing to take care of is getting the carcass plywood broken down. From the ¾" sheet, cut two pieces 13"×25" and two pieces 13"×11" (Figure **B**). Also cut one piece 10⅜"×12" (not shown), to go in the sheet tearing drawer.

**B**

**TIP:** Because the factory edge of plywood tends to get marked up by the time you get it, I like to make my first cuts ¼" wider than I need, and then flip them to cut to final size, assuring four fresh, clean edges.

After orienting the parts for which edges I want to face forward and which sides I want to face out, I stick a Post-it near what will be the top, and draw an arrow pointing to front (Figure **C**). Sure, I could just draw on the plywood, but I like the strong visual cue of a Post-it or tape, and the fewer marks I have to sand out at the end, the more I like it.

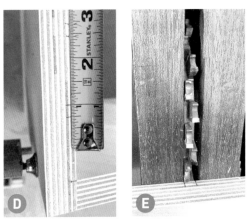

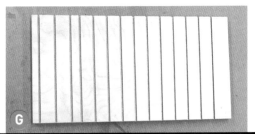

## 2. LAY OUT THE DRAWER DADOS

Next is the layout. In this cabinet there will be 11 drawers for sandpaper above, two deeper drawers at the bottom for scrap sandpaper and for sanding blocks and sticks, and one thinner drawer between them for the sheet tearer.

First I clamp the parts together and start marking out the dado lines. Measuring down from inside the top, make a mark at 1$\frac{7}{16}$" and another at 1$\frac{11}{16}$". Repeat this 10 more times, measuring from the bottom of the previous set of marks (Figure **D**). To help keep track, I make small hashes on the tape measure itself.

The twelfth drawer from the top will be the tearing drawer; for this one I'll mark lines at $\frac{15}{16}$" and 1$\frac{3}{16}$".

To make up for any small discrepancies that happened above, for the last two drawers I measure the remaining distance to the bottom board and divide by 2. This gives me marks at 1$\frac{21}{32}$" and 1$\frac{29}{32}$". Note that the bottom drawer will sit directly on the bottom board.

## 3. CUT THE DADOS

After fitting my table saw with a dado stack at $\frac{1}{4}$" width, and setting it to cut $\frac{1}{4}$" deep (Figure **E**), I make a test cut and check the width of the slot, using a piece of the $\frac{1}{4}$" plywood. It should slide easily with an approximate $\frac{1}{32}$" gap. If it's too tight to slide, you'll need to shim out your dado stack to accommodate. Also verify that your depth is $\frac{1}{4}$" as the rest of the build will rely on it.

Because each plywood piece will be oriented on the table saw to be twice as wide as it is long, it would be very risky to try to cut these dados using the rip fence — a very high risk of kickback, *don't do it*. Instead I use my crosscut fence. Starting with the slot for drawer #1, I align the (not spinning) blade with my marks, then clamp a stop to the fence. Slide the piece back, and then up to the blade again, and recheck that I'm still on my marks (Figure **F**). Once I'm satisfied that my setup is good, I make the cut. Then I take the other carcass piece making sure its top is oriented the same direction, slide it gently to the stop, and make the same cut.

Turn off the saw, and repeat the setup for the rest of the slots (Figure **G**).

**H**

**J**

**TIP:** Because dado sets are typically undersized by $\frac{1}{32}$", there's always a little space between the layout lines and the sides of the blades. For consistency, I try to keep the blade aligned to the same side of the lines for every cut, usually the one closest to the bottom.

### 4. CUT THE RABBETS
When the dados are all cut, it's time to make the rabbets (or "rebates" for the European types). Without changing the height of your setup, change the dado stack to $\frac{3}{4}$" width. The first cuts will be the top and bottom rabbets. On the crosscut slide, I set a stop and remove $\frac{3}{4}$" × $\frac{1}{4}$" deep on both ends of the two side pieces (Figure **H**).

Once those are cut, I move on to the front edge. For the front rabbet, which creates the drawer front recess, I'll use the rip fence and remove $\frac{21}{32}$" ($\frac{5}{8}$" + $\frac{1}{32}$") (Figure **I**).

And once those are cut, I cut a rabbet on the back edge of all four carcass pieces. This rabbet should be $\frac{5}{16}$" wide. And there you have it, all of the drawer slides are cut (Figure **J**)!

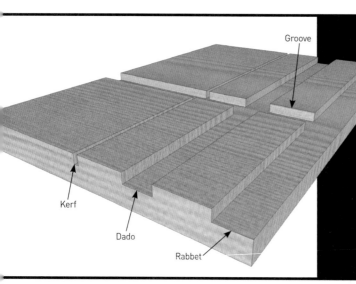

Groove

Kerf

Dado

Rabbet

## Dados and Rabbets?
Yes, this project uses dados and rabbets. Here's a breakdown of the terms:
- A **kerf** is the space left by a saw blade when it cuts, the part that turns into sawdust.
- A **dado** is a wide kerf going *across* the grain.
- A **groove** is a wide kerf going *with* the grain (though these are often used interchangeably).
- A **rabbet** is a two-sided groove (instead of three) cut along the edge of a board.

## 5. BUILD THE CARCASS

At this point I give the whole interior a light sanding, just to remove any splintering around the dados and to smooth things out a little bit.

Because this is shop furniture, I'm going to keep this build as simple as I can using screws and glue to hold it together. If nails or biscuits or finger joints are your thing, I say choose your adventure. Just remember that if you're using fasteners, you will be putting them into the edge grain of plywood, so you'll want to test any nailing on scraps, and predrill for any screws to keep the plywood from delaminating.

I begin by clamping the whole thing together, getting it flush and square, and then making layout marks for my screws. It may seem like overkill, but I find careful spacing of visible fasteners can can give a piece a much cleaner look (Figure **K**).

When everything is laid out I predrill and countersink all my screw holes (Figure **L**), and then disassemble. I apply a light bead of glue to the joints, reclamp, and then drive all the screws. You could glue first and then drill, but I don't like how the drill bit tends to gum up with glue, so I do it this way.

When all the screws are in and the carcass is checked for square, it's time to install the back. After measuring the back opening, I cut a piece of ¼" plywood at 11"×24 1⁄16". Since the ¼" shoulder doesn't leave a lot of room for fasteners, I'll be using #18×¾" nails, driven in at a slight angle, and spaced to avoid the dados (Figures **M**, **N**, and **O**).

## 6. CUT THE DRAWER BOTTOMS

From the remainder of the sheet of ¼" plywood, I cut 14 pieces at 10 15⁄16"×13½" (Figure **P**).

> **TIP:** Were I to do it again, I would cut the plywood with the grain parallel to the short dimension instead of the long. That way the grain would be in the same direction as the drawer fronts, giving it a nicer look.

Slip all the bottoms into their slots and make sure they move smoothly. Then it's on to the drawer fronts.

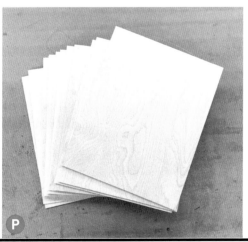

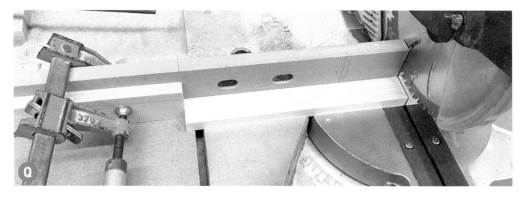

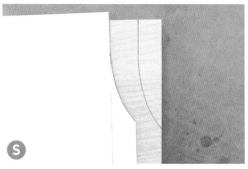

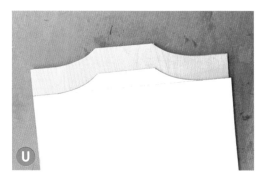

## 7. CUT THE DRAWER FRONTS

To make the drawer fronts, I'm starting with 9 feet of nominal 1×4 maple (true dimensions ¾"×3½") and jointing/planing it down to ⅝" thick. If you don't have access to a jointer and planer, no worries, just select the straightest piece(s) you can and rip them to size on the table saw. Again, to use full ¾"-thick material for the drawer fronts, just make the dimension adjustments listed on page 54.

Next, set a stop on the miter saw to 10¹⁵⁄₁₆" (should match the width of your drawer bottoms), cut two pieces, and rip them to 1¹¹⁄₁₆" wide. These are the two bottom drawer fronts. Then rip the rest of the board into two strips 1⁷⁄₁₆" wide and

cut these into 12 pieces 10¹⁵⁄₁₆" long (Figure **Q**). Lastly, rip one of those down to ⅞" for the tearing drawer.

With the drawer bottoms inserted, I put all the drawer fronts in place and check the fit. There should be about ¹⁄₁₆" space between each drawer, and about ¹⁄₃₂" on each side.

## 8. SHAPE THE HANDLES

The front ⅞" of the drawer bottoms are designed to be used as integral handles. The corners could be left as is, or rounded, or given long miter cuts, or anything else as long as it's only ⅞" deep. I like the look of concave curves to center. To do this, I make a template on a sheet of conveniently sized,

8½"×11" card stock. I measure in ⅞" on the wide end, and then after finding center, I measure out 1⅝" to either side. This gives me a pull that is 3¼" wide.

Because the fit of the drawer fronts relies on that ⅞" mark, to make it easier to land them, I make sure to save part of that mark: I measure in 1½" from the edges as the starting point of my curve. To create the shape, I fool around with some French curves until I find *le sucré* spot (Figure **R**). After cutting out the curves, I trace them to the drawer bottoms (Figure **S**) and cut them out on the bandsaw (Figure **T**). Remember to save your line along the first 1½" on either side (Figure **U**).

When all the handles are cut, I give the edges a very light pass with a piece of 220 sandpaper. Just enough to remove any fuzzy bits, but not enough to round over the corners. Clamp all of them together in a nice square stack and sand the curves to match (Figure **V**), again being careful not to sand past the lines that define the front edge. Once the curves are faired out, I sand each piece front and back, just enough to make it smooth, but not enough to change its shape (Figure **W**).

## 9. BUILD THE BACKSTOPS

Because sandpaper by its nature doesn't slide too much, the drawers don't need much of a backstop, just enough to give the plywood a little stiffness. I'm using 12 strips of ¼" ply ripped at ⅜" wide and cut to 10⅜" long. I cut these on the table saw using a push block (Figures **X** and **Y**), but they could be cut on the miter saw as well.

To attach the backstops, I start by using a square to get them centered (Figure **Z**), then use two pieces of painter's tape to create a hinge along the back (Figure **Aa**). This keeps them from sliding around in the gluing process. When all are taped, one at a time, open the hinge and apply a thin bead of glue along the length (Figure **Bb**). Remember that these will never be under enough stress to break free (unless you're a real drawer slammer), so avoiding a lot of squeeze-out is a good idea.

When they're all glued, clamp flat boards over the ends and allow the glue to cure, usually about

V

W

X

Y

Z

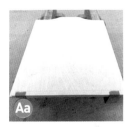
Aa

Bb

**TIP:** To protect against any glue that does squeeze out, I use strips of waxed paper. This might seem like an extra step, but it sure beats discovering that some of your drawers have glued themselves together. So apply the glue, close the hinge, and press flat; repeat and stack them, alternating face to face and end to end, while placing a strip of waxed paper between each one.

Cc

Dd

Ee

an hour (Figure **Cc**).

Of the two remaining bottoms, one is for the bottom-most drawer, the one for sandpaper scraps. For this one I attach a taller piece to the back. I used a 1½"×⅜" cedar scrap I had laying around, but could have used an off-cut of the maple, had I seen that first. The other bottom is for the tearing drawer.

## 10. ATTACH THE DRAWER FRONTS

Once the backstop glue is set, it's time to attach the fronts. Placing one drawer front at the edge of the bench, and a second a few inches back for support, I position a drawer bottom on top of them (backstop strip down). After flushing up the corners, clamp it in place. I'm using 1" drywall screws and glue to connect them (Figures **Dd** and **Ee**), but if you prefer nails, go for it! Note that the ⅞" drawer front should be attached to a drawer bottom *with no backstop*.

## 11. MAKE THE SHEET TEARING DRAWER

The almost last bit of business is to build the sheet tearing drawer. Begin by centering the 10⅜"×12" piece of ¾" plywood in the ⅞" drawer. It should abut the drawer front and be flush with the back. I attached it through the bottom, using a ¾" screw in each corner with no glue, in case I ever want to change it out.

Because I want to be able to quickly tear a sheet in half in long or short direction, I'm going to locate the hacksaw blade 5½" from the left side of the plywood deck. After making reference

Ff

Gg

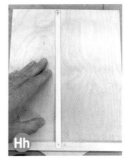

Hh

Ii

Jj

Kk

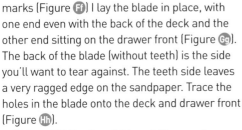

Ll

marks (Figure **Ff**) I lay the blade in place, with one end even with the back of the deck and the other end sitting on the drawer front (Figure **Gg**). The back of the blade (without teeth) is the side you'll want to tear against. The teeth side leaves a very ragged edge on the sandpaper. Trace the holes in the blade onto the deck and drawer front (Figure **Hh**).

Using a ⅝" Forstner bit in a drill press, I carefully drill the front hole until it is flush with the deck (Figure **Ii**). Then using a chisel, I trim the edges of the hole square (Figure **Jj**).

To attach the blade, I use one #6 washer under the blade, and one #6 × ¾" screw and a finish washer, on each end (Figure **Kk**). I predrill and attach the front end first (Figure **Ll**) and the back second, to avoid putting a bow in the blade.

I add two more reference lines, to help me tear the sheets: one at 4¼" from the smooth side of the blade, and then one perpendicular line from side to side. Lastly, since the drawer needs to be taken all the way out to tear lengthwise, I give the back corners of the drawer a good rounding, so that it slots back in easily.

## GET YOUR PAPERS IN ORDER

And there it is! All that's left is to give it all a good sanding, soften all the edges and corners, and apply your favorite finish — but only on the outside, as finishes inside of boxes can go rancid and smell terrible. Or don't finish it at all, letting it bask in its own raw glory.

Then figure out your filing system; high grits up or down? Label the drawers with grit numbers or turn every day into a memory puzzle? The choice is yours. Load it up and enjoy your new and improved organized life! ✐

# The Magic GIF-Ball

## Ask and behold your future told — in memes — with this new twist on a classic toy

**Written and photographed by DJ Harrigan**

**DJ HARRIGAN** is a designer and maker in Northern California. He has built various thingamajigs and written tutorials for Instructables.com, created puzzles for escape rooms, prototyped hardware for SLA 3D printers, and taught people how to solder and operate CNC lasers at makerspaces. He makes videos for Element14 Presents, and for his YouTube channel Mr. Volt, where he builds custom gizmos and animatronics from pop culture.

The classic Magic 8-Ball toy has been a mainstay of pop culture for decades: Ask a yes-or-no question about the future, then turn it over to see your answer "magically appear!" But it's sorely limited by the 20 static replies imprinted on its floating icosahedron. We live in an age of endless stimulation, and equally limitless memes, so why not combine our modern entertainment sensibilities with the familiar form of that classic toy? Why not a Magic GIF-Ball?

I'm not the first person to decide that a fantastical fortune-telling sphere should answer with images instead of text, but I like to think I've made a tidier version that can easily be replicated by anyone with a basic knowledge of electronics and beginner's grasp of the Linux command line. At its surface, this is still a toy, meant to surprise and delight your friends and loved ones as a spin on a familiar object. But it's also an approachable project that tackles inputs, outputs, and programming (if you so choose) for the maker eager for more Raspberry Pi-related goodness.

## BUILD YOUR MAGIC GIF-BALL

Let's do it. Before you build, you might like to watch my overview video at youtube.com/watch?v=wDhnG030C2Q.

### 1. PRINT THE PARTS

You can find the parts packed in a tidy *.zip* over at Element14 (makezine.com/go/element-14-magic-gif-ball). I've included the full assembly as a STEP file in addition to the individual STLs, so you can modify the design to your heart's content.

There are seven parts to 3D print: the upper shell, lower shell, bezel, LCD retainer, mounting block, logo insert, and button retainer. I printed mine out of PLA at a 0.2mm layer height, but at higher resolutions your sphericity will of course be much better. For the shells, I recommend printing in dome vs. bowl orientation as this is less likely to warp. Print the logo vertically (Figure A), as the X-Y axis is much higher resolution than the Z (the logo is a very shallow curved piece, which is a challenge for all FDM 3D printers).

### 2. PAINT

Depending on what color filament you used, you'll need to paint the shell and/or the logo pieces

**TIME REQUIRED:** A Weekend
**DIFFICULTY:** Easy/Intermediate
**COST:** $100–$200

## MATERIALS

- » **Raspberry Pi Model 3 A+ mini computer** Newark 80AC9303, newark.com
- » **LCD display breakout, 1.3", ST7789** Adafruit 4313, adafruit.com
- » **Battery charger, PowerBoost 1000** Adafruit 2465
- » **Push-button power switch breakout** Adafruit 1400
- » **Tactile switch, 6mm, long plunger** Adafruit 1490
- » **Tilt-switch/vibration sensor** Adafruit 1766
- » **LiPo battery, 1800 mAh, single cell** such as Amazon B07TTD2SVC
- » **MicroSD card, 16GB** Adafruit 2820
- » **JST male connector** Adafruit 3814
- » **Jumper wires, female** Newark 42X1200
- » **Right angle headers, male, 0.1", 12 pos**
- » **Machine screws: M3×6mm (4) and M2.5×10mm (6)**
- » **CA glue** aka super glue
- » **3D printed parts** I printed them in black PLA filament, MatterHackers MH Build brand. Download the free 3D files for printing from Element14 makezine.com/go/element-14-magic-gif-ball (free account required).
- » **Spray paint, white**

## TOOLS
- » **3D printer**
- » **Soldering iron**
- » **Paintbrush, fine tip**

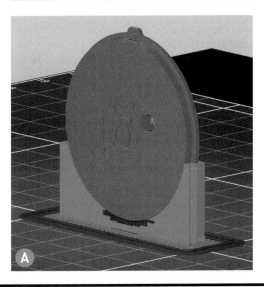

A

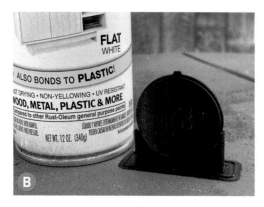

(Figure **B**). If you print the logo in black, you can coat it a solid white and then scrape away (once dry) within the letters to reveal the black underneath, or as I did in the original version, use a fine tip brush to fill in the letters (Figure **C**). I'm not a fan of hand painting, but perhaps you have steadier hands than I!

### 3. SET UP THE RASPBERRY PI

There's very little to do Linux-wise to prepare your Pi (Figure **D**). First, flash the latest Pi OS onto your microSD card (Raspbian Lite is OK as the screen is run directly and doesn't mirror the X framebuffer). Raspberry Pi's new imager tool (raspberrypi.org/software) makes this all much more streamlined if you haven't heard the word.

Get ready for basic headless setup by entering your Wi-Fi credentials. Once connected via SSH, you'll need to install the ST7789 Python module from Pimoroni (github.com/pimoroni/st7789-python) by entering the command:

```
sudo pip install st7789
```

They recommend installing some other common modules first, but I found that these were already up to date in the current OS. Also, be sure to run **sudo raspi-config** and enable I²C and SPI.

### 4. WIRE AND TEST THE DISPLAY

Before connecting any other components, let's test our little TFT LCD display to make sure we've connected it properly and the software is good to go. With the Pi unpowered, use the female jumpers to make the following connections between the LCD and the Pi (Figure **E**); they'll communicate using the SPI serial protocol:

• 3v3 to any 3V pin
• TCS to Pi pin 7 — this is the SPI chip select pin for the TFT display
• SCK to pin 11 — the SPI clock input pin
• SI to pin 10 — the SPI MOSI pin (microcontroller out, serial in) to send data to the display
• D/C to pin 9 — the SPI data or command selector pin
• BL to pin 19
• GND to any Pi ground pin

Double-check your wiring, boot up the Pi, and **cd** into the folder *st7789-python/examples*. Then call:

```
python3 gif.py
```

And you should be greeted by Pimoroni's colorful "Deploy Rainbows" GIF.

## 5. PREPARE YOUR GIF SELECTION

The classic toy is restr...

but ...

Ball...

virtu...

mos...

whe...

If ...

posit...

a me...

class...

The ...

packa...

select...

more ...

accom...

schem...

display...

image t...

as a res...

## 6. SECU...

Once yo...
GIFs, it's ...
to the di...
you've go...
program...
via SCP (S...

```
scp *.
scp fo
```

Your GIFs ...
/home/pi...

```
python3 fortune.py
```

We haven't connected the tilt sensor yet, but you can short BCM pin 2 to GND and the program will register that as a "shake."

## 7. RUNNING THE PROGRAM AT BOOT

There are several options to have a program autorun, but we'll keep it simple. Just call:

```
... /etc/rc.local
```

... exit 0, add:

```
...-c '/usr/bin/python3 /home/
...y' &
```

...mper in half and solder the halves ...he vibration sensor (Figure **F**).
...of 8 right-angle male headers to ...side connections, and a strip of 4 ...s (Figure **G**).
...socket to the pushbutton ...S) (red to IN, black to a GND ..." leads to the pushbutton and ...s 1 and 3. Cut and solder a red

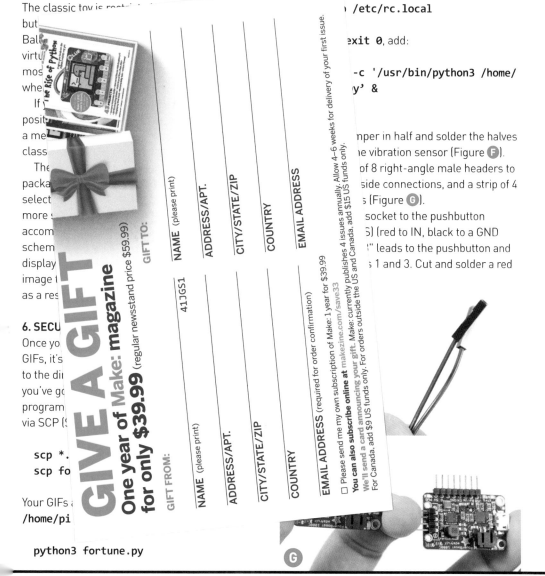

G

jumper to the PPS OUT pin. Cut and solder a black jumper to the G pin of the PPS (Figure **H**).

### 9. ELECTRICAL CONNECTIONS

You already connected the LCD in step 4. Now, connect a jumper from the Pi's 5V pin to the +5V pin of the PowerBoost, and another jumper from a Pi ground pin to a GND pin of the PowerBoost.

Connect the jumper from the PPS OUT pin to the Bat pin of the PowerBoost. Connect the PPS G pin to a GND pin on the PowerBoost.

Lastly, attach one of the sensor pins to pin 2 on the Pi and the remaining pin to a GND pin on the Pi (Figures **I** and **J**).

### 10. PHYSICAL ASSEMBLY

Mount the switch in the square hole within the logo, using two M3 screws (Figure **K**). Glue the logo insert into the lower shell (Figure **L**).

Test-fit the clips in their matching holes, making sure they all seat at the same level when fully inserted. Put a couple drops of super glue on their tips and press them firmly down (Figure **M**).

Screw in the LCD retainer with two M3 screws, then the bezel with two more (Figure **N**).

Place the Raspberry Pi over the four mounting holes and align the mounting block above it, fastening it in place with four M2.5 screws (Figure **O**). Insert the battery (Figure **P**).

Insert the vibration sensor into one of the holes on the side of the mounting block (Figure **Q**). Secure the PowerBoost to the mounting block with two M2.5 screws (Figure **R**).

Wait a few hours for the glue to fully cure, then connect the two halves together, making sure not to pinch any stray wires (Figure **S**).

### NOW ASK THE REAL QUESTIONS

Your Magic GIF-Ball is toggled on and off via the pushbutton period in the .gif logo, which by Pi standards is considered a hard shutdown, so you may want to set it to read-only mode if you're worried about corruption (medium.com/swlh/make-your-raspberry-pi-file-system-read-only-raspbian-buster-c558694de79).

Once the Pi boots up, using the Magic GIF-Ball is very similar to the original toy: Shake it up and receive the answers you so desperately seek! Should I buy Bitcoin? Should I eat fried chicken

**N**

**O**

**P**

**Q**

**R**

**S**

every day? Will I find true love next Monday?

The GIFs are set to loop twice, since most are so short that they usually finish by the time the user is done shaking. Shaking the ball during a currently playing reply won't interrupt it, so if you want that mode of operation you'll have to tweak the code to your liking.

You might also notice that the charging port is not externally accessible, which means it'll have to be popped open to recharge. Is this silly? Yes. But I've shown the MGB to most of my friends and family, and there's still plenty of charge left months later. It's not exactly a daily use fortune-teller anyhow.

I've also pondered a few possible upgrades you may wish to explore:

- **Simplify it.** I went with the Pi as it was the least hassle to prove this as a concept, but this could be adapted to run via an ESP32, Teensy 4.0, or the new Pi Pico!
- **Unlimited GIFs!** It shouldn't be unreasonable

I DON'T THINK SO!

to pull GIFs down via Giphy or another meme-tastic API, but you'll need to make sure your MGB has a constant internet connection.

- **Cleaner power.** Break up the smooth, portless surface with a handy jack for charging, or add another switch to trigger a safe shutdown.
- **Cut corners.** Did you know mini round LCDs are a thing now? Check out Pimoroni PIM570, for example. One of those would make an awesome display that matches the original aesthetic even better.
- **DIY the enclosure.** This project could easily be stuffed into a DIY Christmas ornament, hamster ball, or original 8-Ball.
- **Amp it up.** We've got all this processing power, why not have the MGB respond with full-fledged video clips? Perhaps for version 2... ⊘

Build your own or make a mod? Leave a comment on the original project page: makezine.com/go/element-14-magic-gif-ball

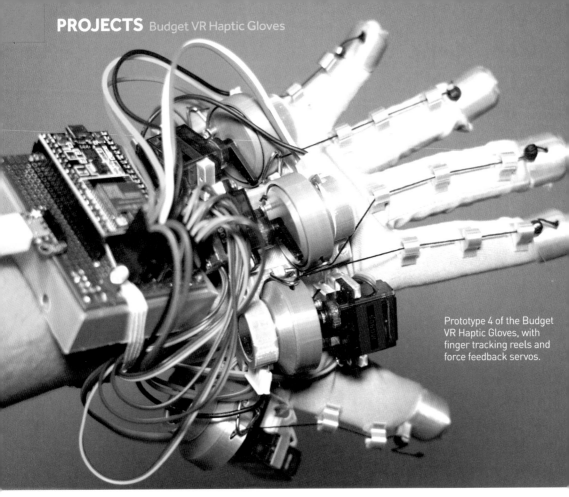

Prototype 4 of the Budget VR Haptic Gloves, with finger tracking reels and force feedback servos.

# Budget VR Haptic Gloves

Written and photographed by Lucas VRTech

## Use your hands in virtual reality and feel the objects you touch — for about 50 bucks

**Stuck at home due to pandemic lockdowns, I turned to virtual reality as a way to keep me occupied during quarantine, but I was very unimpressed by the controllers used to play games in VR.** Not only are they clunky and unintuitive, but they also don't allow you to use your hands like you would in real life: touching,

grasping, and feeling objects.

Solutions do exist, such as commercial VR gloves for professionals, however these cost thousands of dollars and are virtually inaccessible to the average consumer. A few consumer products are available but these can't do force feedback to make it feel like you're really holding

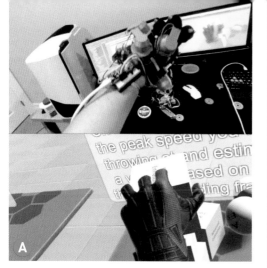

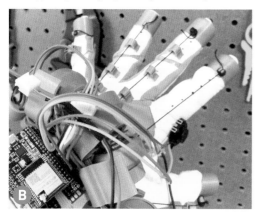

A

Prototype 2, grabbing objects in-game in Unity. It works!

B

Prototype 3, with 5 spring-loaded finger tracking spools.

**TIME REQUIRED:** A Weekend
**DIFFICULTY:** Advanced
**COST:** $20–$60

## MATERIALS
*For each glove:*
» **Potentiometers, 10kΩ, linear taper (5)** such as Amazon B07VQTFFGC or Aliexpress 32869141485
» **Retractable badge reels (5)** used for their rotary springs, such as Amazon B0732Z7T8W
» **Servomotors, 9g (5)** for force feedback haptics
» **Arduino Nano or ESP32 microcontroller board** or clones; use ESP32 for Bluetooth support
» **3D printed parts: spools, tensioners, finger end caps, etc.** Download the free files for printing at github.com/LucidVR/lucidgloves-hardware.
» **A comfortable glove** made of material you can glue onto
» **Elastic** for mounting controllers/trackers
» **Jumper wires, breadboards, JST connectors, etc.** depending on wiring method of choice
» **Batteries (optional)** USB battery bank or 4×AA batteries for wireless operation

## TOOLS
» **3D printer** I use an Ender 3.
» **Hot glue gun**
» **VR-ready PC with any PC-VR headset** Oculus Rift/Quest, Valve Index, HTC Vive, etc.
» **Soldering iron and/or crimping kit (optional)** for more robust wiring

**LUCAS VRTECH** is a computer science student at Massachusetts Institute of Technology, a YouTuber and TikTok creator, and creator of the open-source hardware project LucidVR which aims to make new VR technology more accessible.

things, and they still cost hundreds or thousands.

I really wanted to experience using VR gloves, so I decided to take a shot at building a pair for myself. It started as just a shower-thought, then some napkin drawings, and then eventually I set out to actually build, wire, and program a prototype of my gloves that actually worked (Figure Ⓐ). And instead of thousands of dollars, I was able to build a pair for just $22.

## FINGER TRACKING FOR CHEAP
Some gloves use flex sensors to detect finger position, but these are expensive, about $10 each — that would be $100 or more for the whole hand. Instead, the LucidVR haptic gloves track your fingers using spools of string that are attached to the end of each finger (Figure Ⓑ). As a finger moves, the string gets pulled out of the spool, rotating a potentiometer (Figures Ⓒ

C

Each 3D-printed spool mounts to an ordinary potentiometer shaft.

and 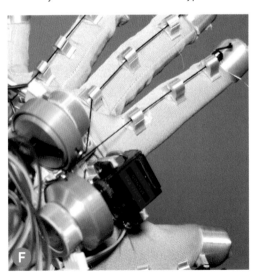), which is measured by an Arduino. The spool is spring loaded, using a rotary spring that's recycled from cheap badge reels, so it retracts automatically. The finger tracking is then calculated and sent to a VR-ready PC over a USB cable or Bluetooth. This allows you to see a representation of your hand in virtual reality that tracks your fingers and lets you interact with objects in the virtual world (Figure **E**).

## PULLBACK FORCE FEEDBACK

Not only can you interact with virtual objects, but now you can feel them too. Prototype 4 of the gloves includes servomotors which pull back on the strings when you hold an object in-game (Figure **F**). This makes it feel like there is actually a solid object in your hand, even allowing you to feel its shape! Adding the servos raises the price a little: expect $20–$30 per hand, depending on where you find your parts. (Some people build only one glove, so they can use a controller in the other hand.) The cost for my current Prototype 4 glove is $23 per hand so far.

All of this software runs off the open source driver called OpenGloves that I have been programming together with a co-developer,

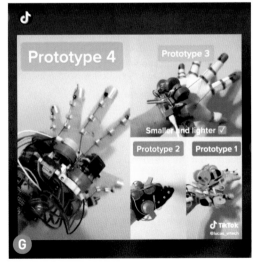

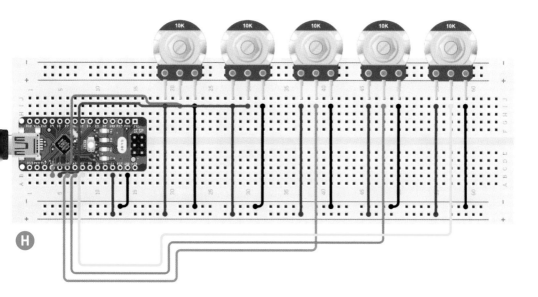

**H**

Danwillm, on GitHub. The OpenGloves driver is now available to download on the Steam store.

## VIRAL VR VIDS

At the same time, I've been publishing videos of my progress on the gloves online in hopes of getting more people involved with the project. To my surprise, one of my very first videos on my VR gloves went viral, with over 8.4 million hits worldwide. Feedback and support started pouring in. It turns out that people enjoy watching my janky-but-functional gloves slowly evolve into a more and more polished product.

Since then, I've been working to add lots of awesome features to the gloves (with hundreds of hours coding them to work). This project has now amassed nearly 400,000 regular followers on TikTok (Figure **G**) and YouTube, and continues to grow every day.

## TOGETHER WE CAN RULE

You can build the Budget VR Haptic Gloves right now and help develop the next prototypes. All the resources for my latest prototypes are available at lucidvrtech.com — the parts list, wiring diagram (Figure **H**), STL files for 3D printing, Arduino software, and our OpenVR driver. Tutorials on YouTube that teach how to build the gloves are available at youtube.com/c/lucasvrtech. And you can get assistance from our LucidVR Discord

community at discord.gg/RjV9T8jN2G.

The gloves are currently compatible with any SteamVR game that supports Valve Index finger tracking, and we are gradually working to roll out mods to add haptics into games.

We're actively looking for contributors to the project. Contributions to the driver and hardware GitHub are always appreciated, and we are searching for more experienced game modders to help integrate our gloves into more games and platforms.

## NEXT LEVEL FEATURES

Force feedback is now working in the game engine on Prototype 4. In Prototype 5, I'll be reducing the bulk of the glove as much as possible (by switching to Hall effect sensors), as well as adding simulation for squishy objects by adding force sensors. A few more planned features we'll experiment with in later prototypes are finger splay tracking, integrated 6-DOF tracking, and vibration haptics.

Feel free to join in, or just follow along with the development of this and project at youtube. com/c/lucasvrtech and tiktok.com/@lucas_vrtech. ◗

# Cellphone Smart Gate

Written and photographed by Mike Senese

## Build this LTE-enabled opener to activate your garage or gate from anywhere

A family member of mine has a basic electric gate at the start of their street that can only be activated by a keypad or standard garage-opener remote. When visitors arrive, the residents have to walk down close enough to use their clicker, or trust telling the guest the access code. Something better would be nice, but a remote-access upgrade would require installers digging trenches to each of the residents' houses and running wiring, or installing a modern but pricey wireless option with monthly data charges.

I've occasionally thought about ways to enable a low-cost, long-range option for them, and was considering assembling a LoRa-based system after we published our articles on that topic in *Make:* Volume 76. Then Particle announced its free cellular-data plan this past spring, offering the foundation for a simple and effective solution.

The ensuing build is a bit Rube Goldberg-esque, but works like a charm: A browser-based web button activates a pin on an LTE-connected Particle Boron microcontroller, causing it to send a current through a connected transistor, which closes the circuit of a dedicated remote garage opener, allowing it to transmit the "open gate" signal. All the pieces are housed inside a basic project box and stored inside the cowling of the gate opener, close enough for the signal to open the gate. The opener mechanism's spare AC outlet supplies the necessary power to keep the board online. Now the gate can be operated from inside the house — or from anywhere else in the world with internet access.

## 1. SET UP THE PARTICLE

Set up your Particle cloud account and the Particle Boron microcontroller per their instructions, using the included antenna (be careful to not crush the antenna socket when inserting it). When it's communicating on cellular data, you'll see the pulsing teal light in the center of the board (Figure Ⓐ).

## 2. PREP THE REMOTE

If you're using a new remote, go to your gate or garage and follow the normal pairing method for it and your opener apparatus.

Back at your workbench, open the remote and remove the circuit board from the housing. On

**TIME REQUIRED:** 1–2 Hours
**DIFFICULTY:** Easy
**COST:** $90

## MATERIALS
» **Particle Boron LTE microcontroller** I used the CAT-M1 version for North America; there's also a CAT-1/3G/2G for Europe and a 2G/3G global version.
» **Garage opener remote control** Be sure to get the right frequency for your gate or garage opener — commonly 315MHz or 390MHz.
» **Breadboard, small**
» **Resistor, 2kΩ**
» **Transistor, NPN** I used a 2N3904; equivalent options include the BC547.
» **Jumper wires, male-male: 3" (2) and 6" (2)**
» **MicroUSB power cord and AC adapter**
» **Project enclosure box** large enough to hold the breadboard, opener, and Boron antenna

## TOOLS
» **Soldering iron and solder**
» **Wire cutters**
» **Wire strippers**
» **Computer with internet connection**
» **Website access** for creating and hosting a new HTML page

**MIKE SENESE** is the executive editor of *Make:*, and a hobbyist problem solver.

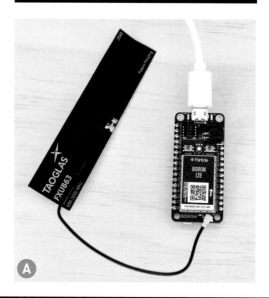
Ⓐ

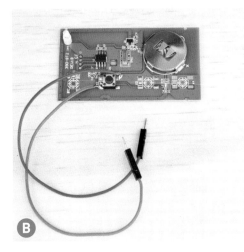

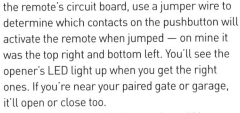

the remote's circuit board, use a jumper wire to determine which contacts on the pushbutton will activate the remote when jumped — on mine it was the top right and bottom left. You'll see the opener's LED light up when you get the right ones. If you're near your paired gate or garage, it'll open or close too.

Next, snip an end off of each of two 6" jumper wires (leaving a "male" end on each one), strip off about ⅛" of the insulation, and solder each one to each of the two contacts you've identified, respectively (Figure **B**).

You'll now have two leads with male pins coming off the remote controller board. Like before, touching these wires together will trigger the remote (Figure **C**) — which is what we're going to make the Boron do for us electronically.

If you put the case back on, you'll need to make an opening for the wire leads you've just added. I left my housing off to camouflage it slightly if someone finds it.

## 3. CREATE THE TRANSISTOR CIRCUIT

We're going to use an NPN transistor as a switch to open and close the circuit on the remote opener. This is a little simpler system than using a relay, which many DIY garage openers use. In basic terms, the two outside leads of the transistor will jump the pushbutton when you apply a voltage through the transistor's middle lead, and thus will transmit the "open" signal.

Put the Particle Boron on a breadboard, toward the top with the pins spanning either side of the center channel (Figure **D**).

In the bottom left corner of the breadboard, one row below the Boron, plug the three leads of the transistor into three rows of the breadboard. I put the flat side of the transistor facing the Boron. Leave an open column to the left of the transistor (Figure **E**).

Plug the remote's new jumper wires into the breadboard, in the holes to the left of the transistor leads (Figure **F**). Put the wire from

**NOTE:** For this version, I chose to leave the remote's battery as its power source; they run for years on a single coin-cell battery, and it's easy to swap it out when the battery eventually dies. A more advanced solution would be to use the voltage from the Boron itself to supply power to the opener — feel free to experiment with that if you please.

the battery side of the remote's button next to the bottom lead of the transistor (the collector), and the wire from the opposite side of the pushbutton to the top transistor lead (the emitter).

In the hole to the right of the bottom lead (the collector) of the transistor (the flat-faced side, as we have it configured), plug in a 3" male-male jumper wire. The other end goes to the GND pin on the Boron (Figure **G**).

Behind the middle pin of the transistor (the base), add a 2kΩ resistor that stretches over the center channel of the breadboard (Figure **H**). You may need to experiment with different values of resistance, depending on what type of transistor you used; too low and the transistor circuit will always be open, too high and it won't open at all. (I narrowed down which resistor to use for my setup by exchanging the remote for an LED in my circuit, setting the Boron's pin to "high", and swapping in resistors of increasing value until the LED light turned off.)

Add a 3" jumper wire between the other side of the resistor and the D8 pin on the right side of the Boron (Figure **I**).

The finished circuit should look like Figure **J**.

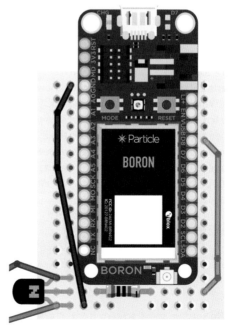

**CREATE NEW APP**

Example apps

1. Blink an LED
2. Web-Connected LED
3. Function Variable
4. Publish
5. Subscribe
6. Tinker

**CAUTION:** Don't share your Device ID and Access Token publicly! They will grant anyone who knows them access to your Particle device. The way we've published it here, a simple look at the HTML page source reveals both of those sensitive elements. If you will share your web app with anyone else, you'll want to utilize a personal server so you can keep the Device ID and Access Token private. Particle has documentation on doing this: docs.particle.io/datasheets/app-notes/an032-calling-api-from-web-page

## 4. BUILD THE SCRIPT AND WEB BUTTON

To remotely tell the Boron to activate the opener, you'll need a script for the board and a web page with a virtual button that accesses it. I modified the code from the Buzz Wolf dog collar project by Nancy Yi Liang (particle.hackster.io/nyl/buzz-wolf-837db0) to make a simple interface for our gate opener.

First, go to makezine.com/go/cellgateopener to download the opener code and web-button HTML (see end box for QR code).

In the Particle IDE at build.particle.io, create a new app for your Boron (Figure **K**). I called mine "gateopener." Paste in the code for the controller and upload to your device.

On the HTML, there are two fields to update in the code for your specific board: **device_id** and **access_token**. You can find your Device ID on Particle's console at console.particle.io/devices.

The Access Token can be a little trickier to locate; the published methods didn't work well for me. What I did to find ours was to click the device link from the same Particle console page, and then click the Terminal tab under the Events section. A popup window will show your Device ID and Access Token (Figure **L**) inside the displayed URL. Copy the token number and paste it into the code above.

Save this document as an HTML file and upload it to your website.

When you navigate to that page on a browser (desktop or mobile, both work the same), you should now see a plain webpage with a blue button that says "CLICK GATE" (Figure **M**) — press that button and a few moments later you'll see the LED on the remote light up. It's working!

**M**

## 5. FINAL STEPS

From here, you can carefully tuck your breadboard into an enclosure (Figure **N**). Be sure to tie a knot into the power cord inside the case so

**EVENTS** View events from a terminal

‖  ▣  🗑  Search for events          ADVANCED

To see the same stream from a terminal, run this:

```
curl https://api.particle.io/v1/devices/████████████/events?acc
ess_token=████████████
```
— Device ID
— Access token

**L**

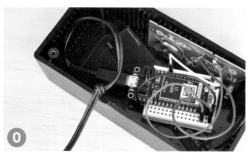

it keeps the cord from getting yanked out (Figure **O**). You may also want to hot-glue your jumper cables and components to the breadboard to help fix them in place.

## SOLUTION DEPLOYED!

Set up your Smart Gate circuit somewhere close enough to your gate or garage for the button press to make it open (Figure **P**). Remember, you'll still need power for the Boron, so look for an accessible AC outlet.

Now you're all set to go — click away and your gate or garage will open, even if you're far away on an overseas vacation.

You can also keep refining this project. A few areas of improvement to consider:
- Powering the remote via the voltage output on the Boron to eliminate battery changes
- Moving everything to a perfboard, or creating a custom PCB for the Boron, components, and remote, to make the circuit more durable
- Adding a sensor to detect and alert that the gate has opened successfully (you might as well use more than just one of those Boron pins, after all)
- Eliminating the opener remote entirely by wiring the Boron directly to the gate mechanism's electronics
- Building an Alexa skill to open the gate with a voice command
- Dedicated Android app, anyone?

There may be more efficient code options as well — if you see ways to optimize it, be our guest.

Mobile garage openers are one of the go-to starter IoT maker projects, but this one brings in the new aspect of near-unlimited range thanks to LTE. Use this project as a basis to think of other LTE integrations, and how you might also incorporate transistor switches into your projects. Have fun! ❂

Grab the code files at makezine. com/go/cellgateopener.

Robert Boyle-inspired vacuum pump

# *Robert Boyle and the*
# Air Pump

Written and photographed by William Gurstelle

**TIME REQUIRED:** A Weekend
**DIFFICULTY:** Easy
**COST:** $100–$200

## MATERIALS

» **Wooden base** See Step 1 below.
» **Wood 2×4, 10" long** for the Guide piece
» **Firm plastic tubing, ⅜" outer diameter (OD):** 24" length and 12" length
» **Three-way valve (diverting valve) with push-to-connect fittings, ⅜" OD connections**
» **Push-to-connect tube fittings (2)** for air and water, ⅜" tube OD × ½" NPT
» **Acrylic bell jar with top opening** such as Eisco Labs, 12" high × 7" diameter
» **Rubber stopper** to fit top opening in bell jar
» **Vacuum gauge**
» **Threaded steel rod, ⅜"-16, 24" long** for the Crank
» **Knob** for ⅜"-16 threaded rod
» **Cylinder, 2" diameter, 2" long** made of Delrin plastic (better) or hardwood (cheaper). I bought my Delrin from McMaster-Carr.
» **Hard plastic sheet, 10"×8"×¼" thick** for the Bell Jar Plate
» **Wood screws, round head, #8×¾" long (4)**
» **PVC pipe, 2" diameter, 16" long**
» **PVC end cap, 2"**
» **PVC square mounting flange** for a 2" Schedule 40 PVC pipe. This is a specialty piece, so you'll likely need to search for it online.
» **PVC cement**
» **O-rings, 2" OD, 1⅝" ID (2)**
» **Tub of grease**

## TOOLS

» Saws for plastic
» Drill and assorted twist bits
» Screwdriver
» Lathe or router, or Dremel with cylindrical grinding bit
» Thread taps, ⅜"-16NC and ½" NPT pipe thread
» Adjustable wrench

**WILLIAM GURSTELLE**'s book series *Remaking History*, based on his *Make:* column of the same name, is available in the Maker Shed, makershed.com.

# Build the vacuum pump that became the symbol of the Scientific Revolution

**The age historians call the Scientific Revolution was an important, pivotal time.** It was the period, chiefly during the 16th and 17th centuries, in which people began to understand nature in modern terms — the astronomy, biology, chemistry, optics, and physics we recognize today — through the work of Galileo, Kepler, Newton, and others.

One of the leading lights of the Scientific Revolution was Robert Boyle (1627–1691), a wealthy Irish aristocrat who is remembered as the first modern chemist. Perhaps his most important contribution was not any single discovery but rather his general influence on science itself. Boyle did a great deal of experimentation and discovery, and his ideas about the scientific method paved the way for generations of scientists who followed.

With his substantial financial resources, Boyle was behind the biggest of all "big science" projects during the Scientific Revolution — the vacuum pump. When you first consider it, it may not seem clear why this was such a terrifically expensive item but at the time, Boyle's air pumps cost a fortune. Every component was designed from scratch and each item made by hand. So important was the air pump that it became emblematic of the Royal Society, the best-known scientific organization of the time. Eventually, the air pump came to represent the entire Scientific Revolution.

Before the vacuum pump, people weren't sure that a vacuum could really exist. Aristotle, the influential Greek philosopher, postulated that "nature abhors a vacuum," meaning that a space without anything in it could not exist. And without a device like Boyle's air pump to actually pull a

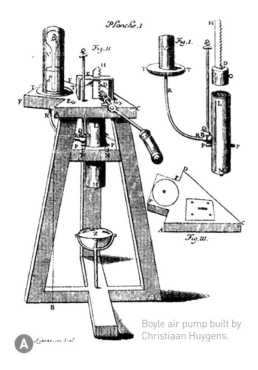

A Boyle air pump built by Christiaan Huygens.

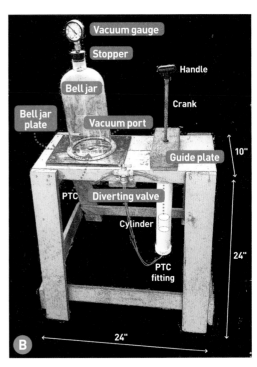

B

vacuum in a space, who could say otherwise?

In an earlier column of Remaking History, we explored the Magdeburg hemispheres, a famous vacuum-related demonstration built by the German experimenter Otto von Guericke (see page 83). Von Guericke is given credit for inventing the very first vacuum pump, but his pump was inefficient and nearly impossible to use for any real scientific purpose.

In order to carry out the important experiments that Boyle wanted to attempt, a better, faster, and more dependable pump had to be invented. Boyle had built for him a series of three different pumps, each one able to generate deeper and longer lasting vacuum conditions than the one before it.

In this column, we will build a replica of Boyle's second pump, originally designed and built for him by Christiaan Huygens (Figure Ⓐ). If built with care, it can pull a fairly deep vacuum, which is useful for all sorts of fun and interesting experiments. Just as importantly, you can get a true appreciation for just how difficult it was for scientists like Boyle, Huygens, von Guericke, and other early experimenters to construct their own scientific instruments.

## BUILD YOUR OWN BOYLE AIR PUMP

During the course of this project, refer to Figure Ⓑ to understand how the components are arranged.

### 1. BUILD A STURDY BASE

Manually pumping the air out of a bell jar is hard work. The piston we'll be using measures 2" in diameter, so the amount of force required to move the piston against near-atmospheric vacuum is in the neighborhood of 45 pounds. That's a lot of force, so a sturdy base is a must.

Figure B shows one way to build a base. You have a lot of latitude here; the important thing is having a surface at least 24" long by 10" wide and 24" high, to hold the bell jar and provide adequate room to operate the pump. I used two 2×6s for the top and 2×4s for the legs and other support pieces, connected with deck screws.

### 2. MAKE A PISTON

Boyle made his pump out of a brass cylinder and a leather-covered wooden piston. But that was before the O-ring was invented. Take it from me, rubber O-rings are *much* better at stopping leaks than leather covers.

Improvised groove cutting fixture.

**C**

Piston with two O-rings, in Delrin (top) or wood (bottom).

**D**

Begin by cutting a square groove, ⅛" deep by ¼" wide, in your piston. The best way to cut the groove is with a lathe. A router is nearly as good if you have a table for it. If you have neither of these items, then you'll have to improvise. I built a simple wooden jig that allowed me to rotate the piston over a vise holding a Dremel tool with a cylindrical grinding head attached (Figure **C**). By carefully rotating the piston I could accurately cut grooves in the Delrin piston.

Once the groove is cut, place the O-ring in the groove and check the fit. There must be about ¹⁄₁₆" of extra width in the groove for the necessary O-ring expansion. In addition, the O-ring will protrude exactly ¹⁄₁₆" above the cylinder. This is a bit more than the clearance between your piston and the interior opening of the 2" pipe, and that's exactly what you want for an O-ring seal. When you insert the piston into the pipe, the O-ring will compress and make an airtight seal against the interior of the pipe.

While one O-ring might be enough to stop any air infiltrating around the piston, a second one adds extra certainty. Cut a second groove as shown in Figure **D**.

Once both grooves are cut, insert both O-rings.

### 3. ATTACH PISTON TO CRANK

Drill a ⁵⁄₁₆" hole in the center of the piston, ½" deep. Use a ⅜"-16NC tap to tap threads in this hole. Screw the ⅜"-16 steel rod into the tapped hole (Figure **E**).

### 4. PREPARE THE CYLINDER

Drill an ¹¹⁄₁₆" hole in the center of the pipe cap. Tap the hole using a ½" NPT pipe tap. Place sealant, or better, an appropriately sized O-ring on the threads of the push-to-connect (PTC) fitting. Use a wrench to turn the PTC fitting into the tapped hole. (This is a prime spot for leaks, so do what you must to stop them!)

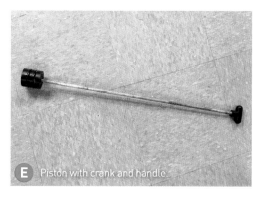

**E** Piston with crank and handle.

**F** End cap with hole and push-to-connect connector installed.

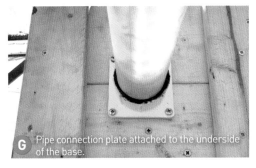

**G** Pipe connection plate attached to the underside of the base.

**H** Bell jar plate with vacuum port, and 3-way valve.

Then, following the directions on the PVC cement can, solvent-weld the pipe cap with the attached PTC fitting to the pipe (Figure **F**). Twist the cap as you attach it to make sure there are no leaks in the solvent-welded connection.

## 5. ATTACH CYLINDER TO BASE
Drill a ¾" diameter hole in the center of the 2×4 guide piece. Next, drill a ¾" through-hole along the centerline of the top of the base, 10" from one side, as shown in Figure A.

Position the hole in the guide piece with the hole in the base so that the two holes exactly align. Attach the guide piece to the base with 4 deck screws.

Now, attach the mounting flange square socket

to the underside of the base. To do so, turn the base upside down, center the socket over the ¾" hole, and attach the socket to the base with four #8×¾" screws, as shown in Figure **G**.

## 6. INSERT PISTON AND MOUNT CYLINDER
Thoroughly grease the O-rings and insert the piston into the cylinder, compressing both sealing O-rings as you do so. Turn the base on its side and insert the exposed end of the threaded rod through the ¾" holes in the guide and base work surface. Then, insert the cylinder into the pipe connection plate. (This connection doesn't have to be airtight.) The ⅜" threaded rod will extend through the hole in the base and guide piece.

Turn the base upright. Screw the knob to the top of the ⅜" threaded rod.

## 7. BELL JAR PLATE
Drill an ¹¹⁄₁₆" hole in the center of the 8"×8" plastic sheet. Tap the hole using a ½" NPT pipe tap. Coat the pipe threads on the push-to-connect fitting with sealant or use an appropriately sized O-ring and then twist the PTC fitting into the threaded hole in the plate using a wrench. (This is a common air leak spot, so seal it well.)

Drill a ¾" through hole in the top of the wooden base for the bell jar port. Place the bell jar plate on the base with the PTC fitting protruding downward into the ¾" hole as shown in Figure **H**. Drill ³⁄₁₆" holes in each corner of the bell jar plate and attach the plate to the base with #8×¾" round head wood screws.

## 8. PLUMB THE PUMP
Attach the three-way valve to the side of base as shown in Figure H. Extend the 12" tube from the bell jar port to the three-way valve. Extend the 24" tube from the cylinder to the three-way valve as shown in Figure B.

To obtain the greatest vacuum inside the bell jar, you'll need to minimize the tube length between the push-to-connect fittings on the cylinder and the exhaust port of the three-way valve. If you want a really deep vacuum, consider moving the three-way valve to the bottom of the cylinder. Placing it there is less convenient, but you'll get a deeper vacuum because there's less dead space in the lines to hold undesired air.

## DRAWING A VACUUM

Thoroughly grease the bell jar lip and place it on the bell jar plate, centered over the vacuum port.

The three-way valve controls the connection between the bell jar, the atmosphere, and the pump cylinder. Before you push the piston down, the valve handle must be positioned so that connection is between the atmosphere and the cylinder. When you pull up, the valve must connect the bell jar to the cylinder.

To draw a vacuum, follow this procedure:

- Start with the piston at the bottom of the cylinder. Just before operating the piston, turn the valve handle so the valve connects the cylinder and the bell jar.
- Pull up on the piston. When the piston is at the top of the cylinder, turn the handle on the three-way valve. This isolates the bell jar and simultaneously makes a connection between the cylinder and the atmosphere, so the air removed from the bell jar can be exhausted when you ...
- Push down on the piston on the return stroke.

With each pull stroke, you're evacuating one cylinder volume of air from the bell jar. The first few pulls will be easy but pulling will become more difficult with each additional stroke as the vacuum inside the bell jar increases.

Continue pumping until you reach the level of vacuum you desire.

## VACUUM GAUGE

To measure the pressure inside the bell jar, add a vacuum gauge. Drill a hole in a rubber stopper, insert the gauge into the hole, and then plug the top hole of the bell jar with the stopper. For tips on using it, see makezine.com/projects/skill-builder-quick-tips-for-using-vacuum-gauges.

## LEAKS AND PATIENCE

Without doubt, the hardest thing about building a Boyle-style vacuum pump is handling the air leaks that inevitably occur. Boyle himself wrote often and bitterly about his problems with leaks. I pretty much guarantee that if you build an air pump, you will have leaks too. Eventually you can stop the leakage, but it will take time.

A prolific writer, Boyle wrote a great deal

## INTO THE VACUUM

Find more cool vacuum projects from the pages of *Make:* at makezine.com/projects.

**MAGDEBURG HEMISPHERES**
Pull a vacuum so strong two horses — or interns — can't separate the halves. The trick? Fire.

**VENTURI-STYLE VACUUM PUMP**
Make a pump based on fluid dynamics and the famous Venturi effect.

**NUCLEAR FUSOR**
Build a high-voltage "star in a jar" to explore nuclear fusion.

**UNIVERSAL ROBOT GRIPPER**
Air pump + coffee + balloon = a "jamming" gripper that molds itself to any object.

about his vacuum-related experiments. Many of these explored the effects of vacuum upon the physiology of small animals, and his notes make for some pretty disturbing reading. Other experiments are friendlier to modern sensibilities and involve the behavior of gases under differing volumes and pressures, the transmission of sound and heat through air and vacuum, and ultimately, the nature of matter itself.

What can you do with your own vacuum pump? There are dozens of interesting projects and experiments, including the propagation of sound through vacuum, heat transfer through vacuum, the respiration of plants, vacuum preservation of foods, and much more. ◉

# Listening to Light

## Build a photophone to send and receive sound on a beam of modulated sunlight

Written and photographed by Forrest M. Mims III

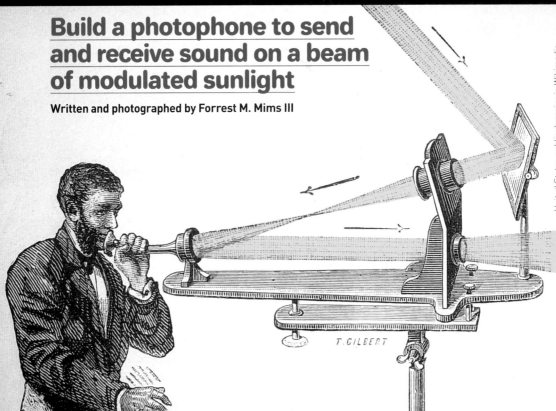

T. GILBERT

United States public domain via Wikimedia

Over the years I've designed many electronic circuits that transform steady (DC) or modulated (AC) beams of light into sound. These circuits have many applications, some of which are described here.

### BASIC LIGHT-TO-TONE RECEIVER

During my senior year at Texas A&M in 1965, I modified a simple 2-transistor oscillator by connecting it to a speaker and replacing its single resistor with a light-sensitive cadmium sulfide (CdS) photoresistor. The circuit was silent when placed at the end of my dormitory's darkened 200-foot hallway. When a fellow student struck a match at the far end of the hall, the circuit emitted a loud buzz that amazed me as much as the students who were watching.

You can repeat this demonstration with the simple audio oscillator circuit shown in Figure Ⓐ. This circuit is easily assembled on a solderless breadboard. It features a 555 timer chip that's connected to a CdS photoresistor across pins 4

## TIME REQUIRED: 2–5 Hours
## DIFFICULTY: Easy
## COST: $15–$50

## MATERIALS
*All parts are widely available online.*
- » Solderless breadboard
- » Hookup wire
- » 9V battery and battery clip
- » CdS photoresistor aka photocell. See text for details; RadioShack offers a set of 5 (item 2761657) for experimenting.
- » Silicon solar cell or photodiode See text.
- » 555 timer IC chip
- » Speaker, 8Ω impedance
- » Capacitors, ceramic: .01µF (2) and 0.1µF (2)
- » Capacitors, electrolytic: 4.7µF (1) and 100µF (1)
- » Resistors: 470Ω (1), 1kΩ (2), 220kΩ (1), and 1MΩ (1)
- » Variable resistor, 10kΩ aka potentiometer
- » LED, red or near-infrared
- » Op-amp IC chip, TLC271 aka operational amplifier
- » Audio amplifier IC chip, LM386
- » Thin mirror or reflective material See text.
- » Bluetooth speaker with clip and phone plug (optional)
- » Battery-powered radio and masking tape
- » Sunglasses and hat

## TOOLS
- » Soldering iron and solder (optional)

**FORREST M. MIMS III**, an amateur scientist and Rolex Award winner, was named by Discover magazine as one of the "50 Best Brains in Science." He has been talking over light beams since 1965. forrestmims.org

and 7. You can slow the circuit's audible frequency range by increasing the value of C1.

This circuit works best with a CdS photoresistor that has a high resistance when dark and a low resistance when illuminated. CdS photoresistors are widely available online. RadioShack offers a set of 5 (item number 2761657) with a wide range of sensitivity differences that allow you to experiment to find the best results.

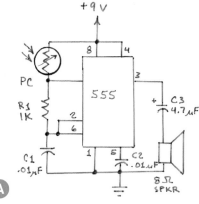

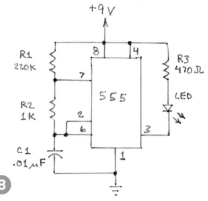

**A** This circuit emits a tone whose frequency is controlled by the intensity of light striking a CdS photoresistor.

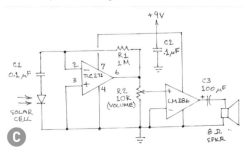

**B** This circuit causes an LED to emit a rapid string of pulses that are converted into an audible tone when the LED is pointed at a light receiver circuit.

**C** This circuit is a light receiver that converts a pulsating light into an audible tone.

## AUDIO FREQUENCY LIGHT TRANSMITTER AND RECEIVER

Figure **B** shows a light transmitter that applies a stream of pulses to an ordinary red or near IR LED at a frequency determined by R1. The values shown provide a light beam that's modulated at about 600Hz.

Figure **C** shows a simple light receiver that will detect the pulsating beam from the LED and convert it into an audio tone. You can use this

A silicon photodiode generates a voltage when illuminated by light.

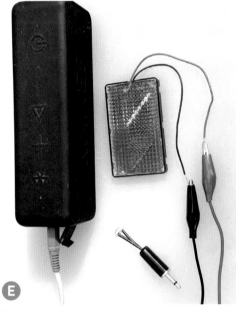

**E**

A light receiver can be easily made by connecting a photodiode or solar cell to the input of a Bluetooth speaker.

**CAUTION:** Be careful if you're connecting a solar panel that has multiple cells or an output exceeding a few volts, for the high voltage might damage your Bluetooth speaker. If you try this, insert a 0.1µF capacitor between the positive solar panel connection and the speaker's positive input, and use the panel indoors, or outdoors only at night.

circuit to test infrared (IR) remote controllers.

You don't need to build the circuit in Figure C if you have a Bluetooth speaker that includes an audio input jack for an external signal. All you'll need to make an audio frequency light receiver is a suitable detector.

I've used a variety of silicon photodiodes for this purpose. A typical one is OSRAM's BPW34. These are miniature solar cells installed in a miniature epoxy or metal package with a glass window. Figure **D** shows the voltage produced by a photodiode in late afternoon sunlight.

Figure **E** shows an Anker Bluetooth speaker connected to a silicon solar cell. The positive lead from the solar cell is connected to the tip of a ⅛" phone plug inserted into the speaker's input jack. The negative lead goes to the ground connection on the plug.

Figure E also shows a photodiode soldered to a phone plug that can be inserted into the speaker's input jack. This works well for testing remote controllers. If no tone is heard from the speaker when the remote's button is pressed, chances are the remote's battery needs replacing.

You can even use an LED as a light-sensitive photodiode instead of a silicon light sensor. Use the same type of LED for both the transmitter and receiver. Near-IR LEDs (880nm) will work best. This is the most common LED used in remote controllers, so you can demonstrate reception of a tone-modulated IR beam simply by pointing a remote controller at the receiver's LED.

Be sure to set the speaker's volume to its lowest level when experimenting. A 0.5-volt silicon solar cell used as a light sensor can cause a very loud sound from the speaker!

## SENDING VOICE OVER A BEAM OF SUNLIGHT

After Alexander Graham Bell invented the telephone, he pioneered ways for beams of sunlight to carry musical tones or even voice. He also pioneered the development of light receivers that converted a tone or voice superimposed onto a beam of sunlight back into sound. He called this new technology the *photophone*.

On February 19, 1980, my wife Minnie and I visited a small parking lot in Washington, D.C. A former building at that site was home to Bell and his wife 100 years before our visit. We were accompanied by representatives from *National Geographic*, the Smithsonian Institution, and Bell Laboratories. Our purpose was to recognize the first time Bell transmitted the human voice over a beam of sunlight by a photophone exactly a century earlier.

To celebrate the photophone centennial, I built a simple version of the device with which we sent our voices across the site over beams of sunlight. We then unwound a 100-foot length of optical fiber to which I had attached a red LED at each end, and we talked to one another with the red LEDs serving as both emitters and detectors of voice-modulated light.

Bell didn't have LEDs or light-sensitive photodiodes. Instead, he made his own light detector from a cylindrical array of selenium cells he cooked on a kitchen stove. When mounted inside a reflector and connected to a telephone receiver and a battery, this simple arrangement formed a receiver capable of detecting the human voice transmitted over a beam of reflected sunlight.

## AN ULTRA-SIMPLE PHOTOPHONE

Bell devised various ways to imprint voice on a sunbeam, the simplest of which was simply a thin mirror. When voice was directed against the mirror, it vibrated in step with the sound waves.

The photophone transmitter I used during the Photophone Centennial was an ultra-thin, 1" circular mirror cemented to one end of a 1"-diameter, 1¼"-long aluminum tube (Figure **F**). Ultra-thin mirrors are hard to find, so an alternate will probably be required. Aluminum foil (shiny side facing out) can be taped to one end of a paper or plastic tube or cup — but the problem with foil is wrinkle prevention.

Aluminized plastic, as from a snack wrapper or a space blanket, is much easier to use. Figure **G** shows a plastic cup with its bottom removed and aluminized plastic taped over the removed bottom.

For your photophone receiver, you can use the circuit in Figure C or a Bluetooth speaker. For best results, use a single, large-area silicon solar cell (e.g., 2"×2" or 4"×4") for a detector. You want to detect only modulated sunlight, so mount the

This photophone transmitter sent the author's voice over a beam of sunlight during a celebration of the Photophone Centennial in 1980.

A simple photophone transmitter made from an aluminized snack package and a plastic cup.

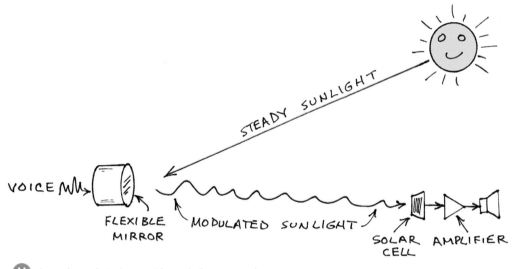

VOICE

STEADY SUNLIGHT

FLEXIBLE MIRROR

MODULATED SUNLIGHT

SOLAR CELL

AMPLIFIER

Arranging a photophone outdoors during a sunny day.

cell inside an open box or oatmeal container to block direct sunlight.

Figure **H** shows how the photophone is used. For initial tests, use masking tape to attach the open end of your photophone transmitter over the speaker of a battery-powered radio. Ask a friend to wear sunglasses and a reflective safety vest and stand or sit in a shaded spot while holding the receiver's detector facing a sunny spot around 50 feet away. Walk to the sunny spot, turn on the transmitter's radio and shine reflected sunlight toward your friend until you see the safety reflectors brighten. Your friend should then hear the radio to which the reflective transmitter is attached.

If a friend isn't available, substitute an inexpensive safety reflector on a pole. Insert the pole in the ground in a shady spot and rotate it so the reflector faces the sunny spot. Next, go to the sunny spot and place the transmitter and radio on the ground or a tripod and tilt it until you see the reflector brightening. Then walk to the reflector and point the receiver's solar panel toward the transmitter.

Either way, be patient. You may need to adjust or remake your transmitter so that its reflector is perfectly flat.

**CAUTION:** While experimenting with a DIY photophone outdoors you should wear dark sunglasses and a hat for protection, and you must never look directly at the sun or its reflection!

## GOING FURTHER

The DIY photophone can be used indoors along dark hallways or inside gyms if the sun is replaced by a bright, battery-powered incandescent spotlight. LED spotlights will not work as well, for silicon solar cells are most sensitive to near-IR wavelengths that white LEDs don't emit. ◑

# 1+2+3

# Virtual Half-Pipe

**Written by Keith Hammond and Mike Senese**

## Turn an old skateboard into a versatile sit-or-stand swing

**TIME REQUIRED:** 1 Hour
**DIFFICULTY:** Easy
**COST:** $10

### MATERIALS
» Skateboard deck, used
» Rope, ⅜" thick, 50' long
» Dowels, wood, 1½" diameter, 1' long (2)
» Hanging hardware (optional) for a tree branch or a beam (see text)

### TOOLS
» Drill and bit, ½" (or large enough to accommodate your rope)

**Here's a great way to repurpose your old beater deck into a swing that's tons of fun and just looks cool.** I saw one at a beach campground near Santa Barbara, California, and immediately decided to make one for my kids. They sell new for $50 to $100 but you can quickly make your own for a few bucks' worth of rope!

## 1. DRILL

Put a hole in each corner of the deck just where it starts to bend up for the nose and tail, 1" from the outside edge. Then drill a hole through each end of the dowels, 1" from the edge as well. (Keep the dowel holes aligned as best you can.)

## 2. KNOT

Cut two ropes to size, then poke each end down through the holes in the deck. Tie a big fat knot into each underneath the deck. Tie another knot into each rope about 30" above, then slide through the dowel handles to keep them sideways over the deck. Tie another knot above each to hold the handles in place.

## 3. HANG

Loop your rope over a tree branch or through your hanging hardware. Hoist the deck to the level you want. Lock in place with an appropriate knot or clip.

You're ready to ride! Be sure to use a branch or beam that's of suitable size and strength. To protect a tree branch from being *girdled* (strangled) by the rope, use short, loose slings made of wide nylon straps or webbing (like car seat belts). Put a carabiner at the apex of your rope, then tighten a U-bolt beneath the biner so your rope's length stays even on both sides forever. ◗

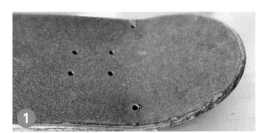

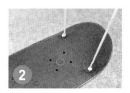

Mike Senese

**KEITH HAMMOND** and **MIKE SENESE** are *Make:* editors and longtime board-sport enthusiasts.

# A Can Full of Tone

Written and photographed by Bob Knetzger

## Build a guitarist's tone toy — in a can!

**Vintage Gibson electric guitars had a special circuit for modifying the sound of the pickups: the Vari Tone** (Figure Ⓐ). You could set the knob to get six different guitar sounds, from a funky nasal tone to a full-bodied sound. Here's an effects box version of that famous circuit for your own guitar that you can build — housed in a hot sauce can.

Why did I use an empty El Pato salsa can? It's just the right size, the metal shields the circuit from radio frequency interference, and it's a nod to guitar amp guru Skip Simmons and his podcast *The Truth About Vintage Amps* (fretboardjournal.com/podcasts/the-truth-about-vintage-amps-big-index-page). Check out his expert recommendations of amps, tubes, and his favorite hot sauce. Listeners have come up with lots clever uses for the iconic El Pato can!

### THE CIRCUIT

Unlike a conventional guitar tone control, this passive circuit uses an inductor coil (Figure Ⓑ) to create a "notch filter" that scoops out selectable frequencies between the bass and the treble. The capacitor/resistor/coil part of the circuit grounds the guitar signal, but just a narrow portion. The frequency is determined by the value of the capacitor wired in series with the coil. A six-position switch chooses one of five different capacitors for five different sounds (Figure Ⓒ). The sixth switch position has no connection, to give an unmodified, bypass sound.

### TIPS

This circuit is great for experimenting! One of the switch positions is wired to a pair of mini alligator clips. You can use this switch setting to quickly test lots of different capacitors (Figure Ⓓ). Plug it in between your amp and guitar and then swap out different values of caps. Listen carefully to each one to find the five capacitors that sound

**TIME REQUIRED:** 1–2 Hours
**DIFFICULTY:** Intermediate
**COST:** $10–$20

### MATERIALS
» **Rotary switch, single pole, six-position**
» **Capacitors (5)** of different values. I used 0.001µf, 0.003µf, 0.03µf, 0.22µf, and 0.33µf — experiment and choose your own values.
» **Resistors, 220kΩ (5)**
» **Audio jacks,** ¼" mono (2)
» **Inductor coil** I used the primary of a small audio transformer, Mouser 42LO21-RC, but you can experiment with whatever you have.
» **Alligator clips, small (2)**
» **Hookup wire**
» **Knob** to fit rotary switch
» **Small tin can** like this El Pato hot sauce can!

### TOOLS
» **Soldering iron and solder**
» **Drill**
» **Step drill bit**

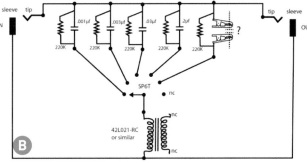

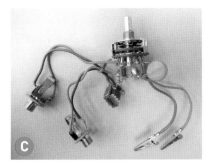

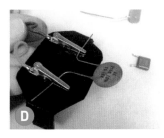

best to you. Then go back and solder those caps into the other switch positions, leaving one position unused for the bypass setting.

To make the holes in the can for the jacks and switch, use a step drill (Figure **E**). It won't grab like a twist drill and you'll get a cleaner hole.

I mounted the mini alligator clips onto a small piece of plastic strip (Figure **F**). Using a soldering iron as a heat source, I softened and bent the sides of the strip to make two ears. Two small magnets, super-glued to the ears, hold the strip in place inside the bottom of the salsa can (Figure **G**).

Now: plug in, "tone" on, and rock out! ⊘

**BOB KNETZGER** is a designer/ inventor/musician whose award-winning toys have been featured on *The Tonight Show*, *Nightline*, and *Good Morning America*. He is the author of *Make: Fun!*, available at makershed.com and fine bookstores.

Larry Chung

Happy
Halloween

Nos Calan
Gaeaf

# House of Fun
## Plug-and-play microcontrollers make home automation easy, at Halloween or anytime

Written and photographed by Helen Leigh

Homebrew alternatives to smart home hubs have been popular in the maker community for a number of years. Lots of tech-aware people are concerned about how Alexa, Google Assistant, and other smart home devices behave when it comes to privacy, advertising, and security issues. Another reason to consider DIY alternatives to these technologies is that it is just straight-up awesome to hack your home with devices you make yourself.

FunHouse is a new board from Adafruit that makes it easy and joyful to make Wi-Fi-connected home automation gadgets, from mailbox sensors to doggie door detectors or cookie jar burglar alarms. FunHouse has built-in sensors for temperature, humidity, light, and pressure, plus three Stemma connectors and a Stemma QT port for I²C devices. The libraries are great too, letting you easily connect your projects with home automation software such as Home Assistant.

For this make, we'll take a look at some of the things the FunHouse can do by making a haunted house that senses movement, then dispenses jelly beans and lights up a window filled with spooky shadows. We'll also connect our project to the internet and log temperature data from the FunHouse using an Adafruit IO account. When you've completed this project you'll see how easy it is to automate your own haunts, props, or household devices, any day of the year.

## BUILD A MINI HAUNTED HOUSE

### 1. CONNECT YOUR MOTION SENSOR
Start off by connecting the PIR motion sensor to the FunHouse board. A passive infrared (PIR) sensor is made up of pyroelectric sensors that detect heat energy. You have almost certainly interacted with a PIR sensor before by walking through doors that open automatically or waving your arms to turn on motion-triggered lights.

Take a look at the very top of your FunHouse board and you'll see a black circle marked PIR, with some pins and symbols underneath it. Next, look at your mini breadboard-friendly PIR sensor and you should notice the labels VIN, OUT, and GND next to the three pins. Plug your PIR sensor into your FunHouse board, matching the VIN pin with the hole labeled "+" and the GND pin with the hole labeled "–" (Figure A).

**TIME REQUIRED:** 2–3 Hours
**DIFFICULTY:** Intermediate
**COST:** $60–$70

## MATERIALS
» **Adafruit FunHouse Wi-Fi Home Automation microcontroller board** Adafruit 4985, $35
» **Mini PIR motion sensor** breadboard-friendly, Adafruit 4871, $4
» **NeoPixel LED strip** with Stemma connector, Adafruit 3919, $13
» **Servomotor, micro** with Stemma connector, Adafruit 4326, $6
» **Craft paper, black**
» **Tape and glue**

## TOOLS
» **Scissors**
» **Computer with internet connection**
» **Code and files:** github.com/helenleigh/funhouse

**HELEN LEIGH** is a hardware hacker who specializes in music technologies and craft-based electronics. Say hello to her on Twitter @helenleigh.

**A**

### 2. CONNECT YOUR OUTPUTS
On the side of the FunHouse board you will find three connectors labeled A0, A1, and A2. These are Stemma-format 3-pin JST connectors for connecting NeoPixel strips, servos, speakers, or relays. These pins can function as analog inputs or digital I/O. Above these you'll find another connector labeled I²C, which is a 4-pin JST that allows you to connect Stemma QT and Qwiic inputs and outputs, or Grove I²C devices if you get a $2 converter cable.

Connect your servomotor with Stemma

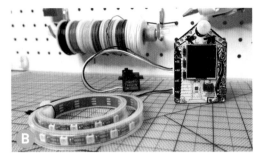

connector to pin A0. Next, connect your LED strip with Stemma connector to pin A2 (Figure **B**). That's all your wiring done! Next up: code.

## 3. SET UP CIRCUITPYTHON

The easiest way to code your FunHouse is to use CircuitPython. Go to circuitpython.org/board/adafruit_funhouse and download the UF2 file. Make a note of the release number, as you may need it later when choosing which libraries to install. Connect your FunHouse to your computer using a data/sync cable. Cables meant for charging only will not work, so if you're having trouble connecting to your FunHouse, check your cable!

Click the reset button next to the USB-C port on the FunHouse, then about a half second later, while the lights on the board are purple, click the reset button a second time. When you get your slow double click timed right, a disk drive called *HOUSEBOOT* will appear. Copy the downloaded UF2 file onto the *HOUSEBOOT* drive and your FunHouse should automatically reset. *HOUSEBOOT* will disappear and a *CIRCUITPY* drive will appear.

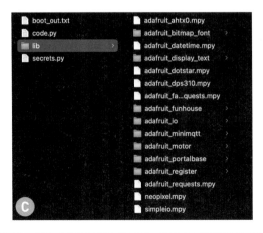

## 4. SET UP CIRCUITPYTHON LIBRARIES

To help us use all the different features of the FunHouse, we need to download and install some libraries. Go to circuitpython.org/libraries then download and unzip the folder of libraries. Take a look inside the folder and you'll see a bunch of folders and files for all sorts of purposes, from Bluetooth to GPS to environment sensors. The whole folder of libraries is too big to fit onto the FunHouse, so you need to choose the ones you need for your project.

For this project you'll need the libraries shown in Figure **C**, so find them all and copy them into your *CIRCUITPY* drive, inside the */lib* folder. (The library list is also available at github.com/helenleigh/funhouse.)

## 5. SET UP AN ADAFRUIT IO ACCOUNT

When you make projects that connect to the internet, you'll often need to fetch and use the current date and time. The easiest way to get this information is to create a free Adafruit IO account so you can use their time service. Adafruit IO is a secure platform that lets you inspect and interact with your device's data.

Go to accounts.adafruit.com to register and make an account. Once you have your account set up, log into io.adafruit.com, then click on My Key in the top menu bar. Make a note of the username and key that show in the pop-up box. Do not share these details with others.

Go to the Feeds tab in the menu bar and select View All. Click the Create a New Feed button and create a new feed with the name **temperature**.

## 6. CREATE A *SECRETS.PY* FILE

To access the internet your FunHouse will need to know your Wi-Fi and Adafruit IO information. The main body of your code will live in a *code.py* file on your *CIRCUITPY* drive, but you should always keep your private information (like your Wi-Fi details and logins) in a separate file so you don't accidentally share your secrets when you share your main code. Never publish your *secrets.py* file.

Open up your code editor (Mu Editor is recommended for CircuitPython) and create a new file. Inside that file you'll need to include your Wi-Fi network name and password, plus your Adafruit IO username and key, along with your time zone. Here's what your code should look like (Figure ) but with your details inside the quote marks. To find your correct time zone, go to worldtimeapi.org/timezones. If your location is not listed, look for a city in the same time zone. Call your file *secrets.py* and save it to your *CIRCUITPY* drive.

## 7. TEST YOUR INTERNET CONNECTION

This bit of code, *iotest.py*, is designed to test that your internet and Adafruit IO account are working correctly (Figure **E**). In lines 11 to 13, the function **log_data** connects to the network and pushes data from the onboard temperature sensor to the **temperature** feed. I don't want my FunHouse to push data to Adafruit IO too often, so on line 9 I checked for the current local time and date, then stored that information in a variable called **timestamp**. Line 18 checks again for the current local time and date and then subtracts the information contained in **timestamp**. If the difference between the two is greater than 30 seconds, it will set **timestamp** to the current local time and date then call the **log_data** function to send your data to Adafruit IO.

Open up your code editor again and create a new file containing this test code. You can type it out or copy and paste it; you can find all the test code at github.com/helenleigh/funhouse. Save this file onto your *CIRCUITPY* drive as *code.py* and it will start running.

> **WARNING:** If you already have a *code.py* file on your *CIRCUITPY* drive you will save over it, so make sure you back up any code before saving over it.

After 30 seconds check your temperature feed and you should see your first data point! For your own code, you probably don't need to check the temperature every 30 seconds.

You should also spend some time exploring what Adafruit IO can do for your connected projects. To start, try creating feeds for the other sensors on your FunHouse then setting up a dashboard to visualize your data and interact with your board.

## 8. ACTIVATE YOUR SENSOR AND OUTPUTS

This code, *inputsoutputstest.py*, tries out the PIR sensor, the servomotor, and the NeoPixel strip (Figure **F**). Lines 1 through 7 pull in the libraries you need and lines 9 through 14 set up your outputs. If you've attached your servo to pin A1 instead of A0, or if you want to turn up the brightness of your NeoPixels, this is where you'd do it.

The next chunk of code creates a function called **color_chase** that makes the strip of NeoPixels fill with color, one after another. After that, we define the RGB value of each color. In the **while True** loop, if the PIR sensor detects motion it will move the servo, light up the

iotest.py  inputsoutputstest.py  displaytest.py  code.py

```
 1  import time
 2  import board
 3  from adafruit_funhouse import FunHouse
 4
 5  funhouse = FunHouse(default_bg=0x000F20, scale=3)
 6
 7  funhouse.display.show(None)
 8  text_row_one = funhouse.add_text(text="Happy", text_position=(12, 18), text_color=0x606060)
 9  text_row_two = funhouse.add_text(text="Halloween", text_position=(12, 30), text_color=0x606060)
10  text_row_three = funhouse.add_text(text="Nos Calan", text_position=(12, 48), text_color=0x606060)
11  text_row_four = funhouse.add_text(text="Gaeaf", text_position=(12, 60), text_color=0x606060)
12  funhouse.display.show(funhouse.splash)
13
14  while True:
15      print("hello!")
16      time.sleep(30)
```

**G**

NeoPixels, and move the servo again.

Open up your code editor and create another new file containing the *inputsoutputstest.py* code. Save this file onto your *CIRCUITPY* drive as *code.py* to run it, again making sure to back up any code before saving over the previous file. Wave your hands around in front of the PIR sensor and you should see your NeoPixels light up and your servo waving.

## 9. TRY OUT YOUR DISPLAY

The FunHouse board has a built-in display that will come in handy for lots of home automation projects. You can display sensor readings, information from Adafruit IO, text, or even shapes and bitmap images. For this make I kept it simple (Figure **G**), displaying four lines of text reading "Happy Halloween" and "Nos Calan Gaeaf," which is an equivalent tradition in my home country of Wales. Again, copy the code *displaytest.py* and save it as *code.py* to test it out.

## 10. PUT ALL THE CODE TOGETHER

When I'm writing code for a project with a few different parts, I like to test smaller chunks of code separately to help me troubleshoot things at each stage. Now that you've made the internet, display, and inputs/outputs work, you can combine all the code to make your final *code.py* file. Or grab my main *code.py* file at github.com/helenleigh/funhouse.

## 11. BUILD YOUR HOUSE

With your electronics and code working, you can move on to the construction of your Halloween FunHouse. By layering black card and semi-translucent paper you can make a cool and creepy shadow puppet-style effect when the LEDs light up. Using craft card, cut out a house shape with a space for the FunHouse board, a space for the servo, and a large window for the shadows to show through. Then cut some shapes to cast spooky shadows. I chose tentacles but you should choose whatever you want: zombies, ghouls, witches, spider webs or all of the above! You'll also need to add wings on either side of your house so you can fold them back to keep the house upright (Figure **H**).

Layer translucent paper over your window frame and secure it in place with tape. Next, arrange your spooky silhouettes and tape them in place too (Figure **I**). To check that you're getting the effect you want while you're placing and positioning your silhouettes, shine a flashlight from the back to the front.

**H**    **I**

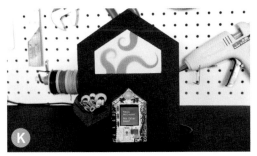

## 12. MAKE A GOOGLY EYE (OR JELLY BEAN) DISPENSER

Push your servo into place on the wall of your Halloween FunHouse, securing it with tape if needed. Next, cut out one circle of card and one tabbed semicircle. Fold the tabbed semicircle in half, then secure it to the circle at about a 90 degree angle (Figure **J**). Use glue to attach a servo arm to the other face of the circle, then attach the whole thing to the servo to make a rotating dispenser.

## 13. FINISH YOUR HALLOWEEN FUNHOUSE

When your googly eye dispenser is in place, you can assemble the rest of your Halloween FunHouse. Fold back the wings on either side to keep the structure upright, then stand the FunHouse in the doorway (Figure **K**). You can make the FunHouse board stand upright by screwing in some M3 screws at the base and you can keep the light out of the cracks with some black masking tape.

To put the finishing touches to your project, you can customize the look of your Halloween FunHouse by making different house styles with different creatures or experimenting with casting the shadows and diffusing the LEDs through different materials.

## LEVEL UP

Now that you know how to code and connect the FunHouse, what else will you automate? Share your builds and ideas at makeprojects.com. ◎

# Smart 'n' Private Home Projects

• **Dr. Anuradha Reddy** @anu1905 is a postdoctoral researcher in hacking and IoT at Malmö University who likes to combine textiles and hardware. She wanted her pillow to tell her when it was cozy warm, so she installed a temperature sensor and hacked a lamp to display the temp through changes in the ambient light, using an Arduino and a Raspberry Pi hosting Mozilla WebThings. anuradhareddy.com

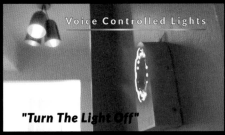

• **Jen Fox** @jenfoxbot is an engineer, maker, and educator who works at Microsoft and builds all sorts of things on the side. Jen's dog Marley likes to trap himself in rooms while she's gone, so she used a Micro:bit microcontroller to make this awesome, dog-friendly automatic door that opens when Marley presses the switch. It's renter- and beginner-friendly, with no prior experience necessary. foxbotindustries.com/dog-door-opener

Voice Controlled Lights

*"Turn The Light Off"*

• **Alfred Gonzalez** @Alfred_G_C wanted to avoid nosy services like Google Assistant and Alexa but still get the feeling that he lives in the future. So, in a futuristic effort to avoid having to get up off his couch, he made this voice-controlled light switch with colleague **Samreen Islam**, using their MATRIX Voice board, a Raspberry Pi, a relay, and Rhasspy — an open source, offline, private, voice assistant with a simple web interface for training your own words and commands. hackster.io/matrix-labs/projects

# Wicked Wings

## Rise above the rest by adding articulating wings to your next costume

Written by Mike Senese

Demon hunter cosplay from *World of Warcraft*, by Willow Creative

**MIKE SENESE** is the executive editor of *Make:*

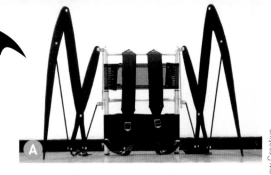

**Flight has caught our fancy since the dawn of time.** Ancient Greek myths detailed Daedalus strapping wax-and-feather wings to his arms. Leonardo da Vinci diagrammed wearable winged mechanisms in the 15th century. Otto Lilienthal helped launch the sport of hang gliding at the end of the 1800s, and modern wingsuits have been around since the mid-1990s. This is to say, humans have always wanted wings.

Some members of the costume community feel the same way, and have built their own ornamental winged accessories. These always capture great attention due to size, realistic articulation, and that basic human desire to flap ourselves into the sky.

Putting articulating wings onto a costume can be a big undertaking. You'll need a harness, an inner framework for the appendages, a method of motion (be it manual, electric, or otherwise), and outer materials for the visual aspect. The following wing projects will help get you started; all use a simple principle of geometry where the angles of a parallelogram change when its shape is skewed. Push a side in one direction and an attached appendage will stretch outward. Pull the side back in and the pieces fold back together.

A.Z.Production (@azproductioncosp, Rachel McConnell, Ted Bruyere, Willow Creative

## Ⓐ MANUALLY ARTICULATING WINGS

**By Rachel McConnell** instructables.com/Articulated-Wing-Framework

This is a good starting point for a build. Rachel McConnell uses lightweight balsa wood for her wing framework, and attaches them to the wearer's arms. Raise and lower your hands to make the wings open and close, and pull your arms forward for a flapping motion. "Suitable for angels, bats, demons, birds (but not so much for butterflies or other insects)," she writes. "Decorate them with leather, feathers, streamers, gold paint, whatever strikes your fancy. Mine ended up with a 9-foot span when open fully."

## Ⓑ PNEUMATIC WINGS

**By Ted Bruyere** instructables.com/How-to-build-pneumatic-costume-wings

Once you've got a moveable set of wings, you might want to give them a power source. Creator Ted Bruyere uses a pneumatic 4-inch throw piston to apply force onto the square aluminum

tube wing frames, expanding them open. The wings then fold back in when the piston compresses. A canister of $CO_2$ supplies power, tucking into the backpack that the wings mount to. Bruyere strongly advises to limit the pressure on the system — the pneumatics will pop the wings open extremely quickly at full power.

## Ⓒ ELECTRIC WINGS

**By Willow Creative** willow-creative.nl/build-logs-tutorials/build-log-animatronic-wings

Using a more optimized parallelogram-frame construction, Willow Creative's approach substitutes an electric linear actuator for manual or pneumatic power. Batteries and control switches are routed strategically through the costume. The arms themselves are made from PVC, using custom 3D-printed linkages.

Willow Creative also includes tips on finishing the wings, both as leathery demon-styled and feathery angel. Her work is top notch, and it's those finishing touches that take a costume to the next level, so pay close attention. Ⓞ

Written by Jaimie and Jay Grenier, Wicked Makers

# Little Props and Horrors!

## Tis the season... to spook the neighborhood

**Halloween is our favorite time of year!** It's a perfect holiday for makers because we get to stretch our creative muscles and make some really fun props and decorations that we can share with our entire community throughout the season. It's a time when anything goes and you can show off your skills. Whether your theme is scary, silly, funny, magical, whimsical, or high tech, there are no rules for a DIY Halloween and your neighborhood kids are sure to enjoy it no matter what you build! Here's a roundup of some of our favorite haunted DIY projects to inspire you to make your Halloween more magical:

### Ⓐ BUBBLING WITCH'S CAULDRON

youtube.com/watch?v=GKuTqjnWoGI

The Witch's Cauldron is a classic Halloween prop and incredibly fun to make! Elevate a store-bought cauldron by painting on a rusty finish. Mix ModPodge with oatmeal and paint it green to create a chunky, overflowing witch's brew. Burning embers are easily simulated by spraying expanding foam over orange string lights. Use small ultrasonic pond foggers to create bubbles and mist rising from inside. A tripod made from lightweight bamboo poles is perfect for hanging your cauldron above your glowing coals.

## B HALLOWEEN TOMBSTONES

youtube.com/watch?v=vEpFd1oqwDA

No Halloween haunt is complete without a creepy cemetery! To create your own custom tombstones, start with XPS insulation foam and carve your shape and an epitaph with a rotary tool or a utility knife. Add an instant faux stone texture by spraying it with water and running over it with a heat gun. Seal it with Drylok for a sturdy, weather-proof finish and paint it with gray, green, brown, and black acrylic washes.

## C CREEPY WALLPAPER STENCIL

youtube.com/watch?v=pKjmqxmavFc

Going for the haunted Victorian mansion look? This genius hack by **VanOaks Props** (instagram. com/vanoaksprops) will let you quickly re-create the look of antique wallpaper without the tedious work of moving and aligning a traditional stencil. The secret? Use a lace curtain! This and a spray gun will let you paint an entire 4×8 wall panel in just a few minutes.

## D GIANT JACK-O-LANTERNS

youtube.com/watch?v=nkMa8sqXTlA

Create a Halloween display straight out of *Alice in Wonderland* with these giant whimsical jack-o-lanterns by **Kara Walker Designs** (instagram. com/kara_walker_designs). Starting with a giant block of foam, carve your rough shape with hot wire tools and then sand it smooth. Cut it in half to hollow it out and carve in the face. After gluing it back together, sculpt a twirly stem and some high points around the eyes, nose, and mouth to give it that extra spooky character. Hard coat it to make it durable and finish it with paint.

## E MONSTER IN A BOX

instructables.com/Monster-in-a-Box-2

This wildly fun prop from **Craig Jameson** uses an Arduino and an old car wiper motor to create the illusion of a scary monster trapped inside a wooden crate! The Arduino and motor turn a cam that shakes the top of the box, while LEDs and a fog machine create the illusion of something sinister hiding inside. In this detailed Instructable, Craig provides the schematics for wiring everything up as well as the Arduino code to create your own Monster in a Box. ✪

WICKED MAKERS

**JAIMIE GRENIER** has a background in Hollywood special effects, and **JAY GRENIER** works in the VFX/animation industry. You can find more of their Halloween DIY projects at youtube.com/wickedmakers.

# GENERATIVE
# DESIGN

## Use the power of AI to build biomimetic, optimized models within Fusion 360

### Written by Lydia Sloan Cline

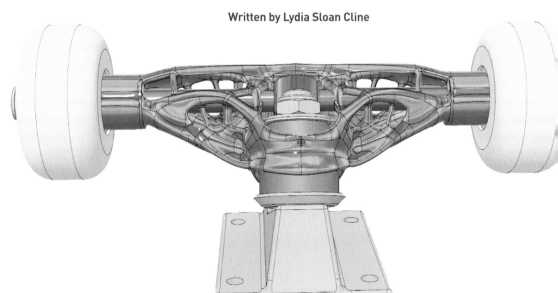

**LYDIA SLOAN CLINE** teaches Fusion 360, 3D printing, SketchUp, and board drafting at Johnson County Community College in Overland Park, Kansas. Her previous works include *3D Printing with Autodesk 123D, Tinkercad, and MakerBot; 3D Printing and CNC Fabrication with SketchUp; 3D Printing Projects for Makerspaces;* and *Architectural Drafting for Interior Designers, 3rd Edition.*

A utodesk recently added a futuristic function to its Fusion 360 design software. The new Generative Design workspace uses cloud-based artificial intelligence (AI) software that designs mesh and T-spline models with parameters that you provide. It's called *generative* because the software generates dozens or hundreds of solutions from those parameters. If you don't like any of the results, you can input different parameters. The software will offer different, improved outcomes via *machine learning*, the ability to learn from and analyze data without being specifically programmed to do so.

Generative design is not exclusive to Fusion 360. Autodesk Revit, PTC's Creo, and Siemens NX are other programs that have it. This software has produced buildings, products, and construction materials.

## WHY USE GENERATIVE DESIGN?

Generative design can produce lighter, stronger, more cost-effective, and aesthetically pleasing outcomes than you could think up on your own. It frees you from your personal limits on imagination, time, history, engineering biases and experience. Outcomes often look different from traditional solutions, such as the wheel rims in Figure **A** and the airplane partition in Figure **B**. This is because the AI uses biomimicry and evolutionary algorithms to produce designs that are organic looking. You're unlikely to get a perfect, finished design from the generative process, but you'll get a base design to edit and develop.

## ACCESS THE GENERATIVE DESIGN WORKSPACE

Find it in the workspace switcher of a paid, educational, or 30-day trial version of Fusion 360 (Figure **C**). Depending on your version, you may need cloud credits to run it. The trial version is limited to 300 cloud credits. Buy them through

Wheel rims for a Volkswagen Microbus.

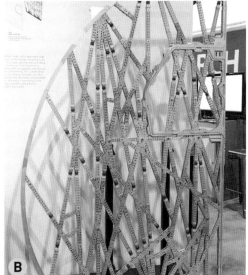

An Airbus partition that was generatively designed and 3D printed with Scalmalloy, a material that was also generatively designed.

The Generative Design tab.

Autodesk, Lydia Sloan Cline

**SCALMALLOY** is a high-performance alloy metal made from scandium, aluminum, and magnesium. Lightweight, strong, and ductile, it was developed by Airbus specifically to 3D print parts for airplanes.

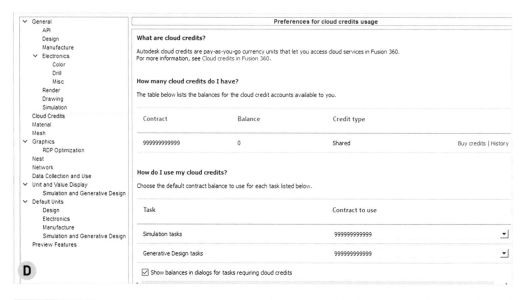

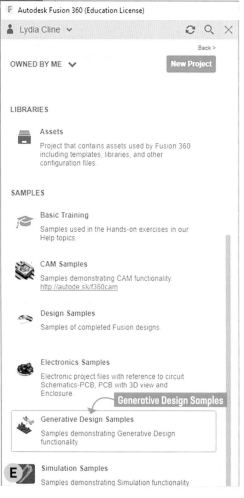

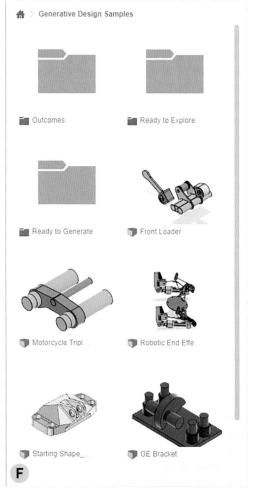

the Preferences panel: General → Cloud Credits (Figure **D**). (Note: This system appears to be changing — check Autodesk's website for the latest details). Generative design is not available in Fusion's free version.

## ACCESS IN-HOUSE LESSONS

Open the data panel. If it shows your design files, click on the house icon in the upper left to access the project list and scroll down to Generative Design Samples (Figure **E**). Click it open for projects to experiment with (Figure **F**). I clicked on the GE Bracket (this particular sample may not be available in all versions). All these files are locked, so right-click on the Browser root and choose Save Copy As (Figure **G**). Scroll back up the project list to Your Corner and find the copy there. Click it open, and then click on the Generative Design workspace.

## THE INTERFACE

The Generative Design workspace interface looks similar to all the other workspaces, except that it shows a milling (cutting) tool at the origin point (Figure **H**).

**G**

**H**

**I**

The Browser has fields for inputting the information required to generate a study. There are also listings for any studies already done. You can right-click and rename past studies from the Browser, or just delete them. Each study listing contains all its inputs.

Click on the Ribbon's icons from left to right to input your information in the correct order (Figure **I**). The first icon, Guide, opens a learning panel on the right side of the screen (Figure **J**).

Now let's model something more relatable than the GE Bracket to show how generative design works.

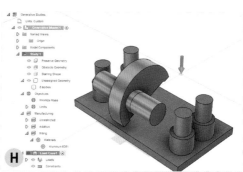

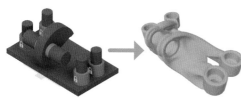

**J**

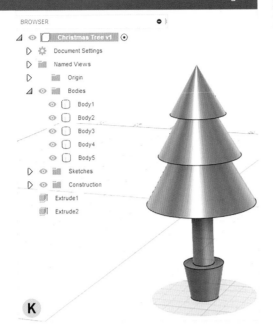

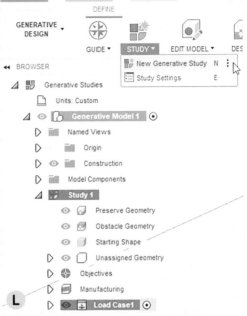

## GENERATIVELY DESIGN A TREE

Model a part in the Solid workspace to give the AI something to work with. It can be a single body, multiple bodies, or a component assembly. It can be very basic, maybe something that just has mounting holes, flanges, or parts that will carry loads. Remove any non-essential parts like fasteners, pins, fillets, and chamfers. Turn off the visibility of all parts that you don't want considered in the study.

Figure **K** shows a tree modelled in the Solid workspace. It consists of five simple bodies: three cones, a straight cylinder, and a tapered cylinder. Bring it into the Generative Design workspace and we'll go through the icons from left to right.

### STUDY

In the Generative workspace, click on the New Generative Study icon to put a "Study1" entry in the browser (Figure **L**). Everything you input will go here. Then right-click on the study. This brings up a window that lets you choose the result's resolution: Coarse will generate faster than Fine (Figure **M**). You might want to generate several coarse studies to save your cloud credits before generating a fine one.

### EDIT MODEL

This lets you create additional bodies to serve

as *obstacle*, *preserve*, and *starting* geometries (Figure **N**). Those geometries don't have to be created in the original design.

Click to access Solid and Surface submenus and run the mouse over the icons for tooltips. In the Modify menu, the red X removes features, and the blue dot replaces selected objects with primitives. Know that the changes you make in the Generative Design workspace are permanent, not temporary. The Edit Model function is there so that you don't have to leave this workspace to go into other ones.

Let's remove the lower two cones by selecting them in the Browser and deleting (Figure **O**). Then click Finish Edit Model.

### DESIGN SPACE

Choose what to preserve and what to serve as obstacles (Figure **P**). Preserved geometry are items to remain untouched. Obstacle geometry are areas in the design where we don't want the

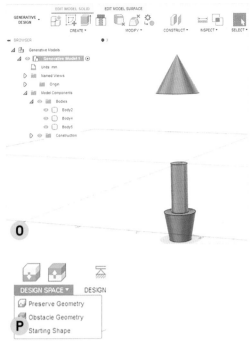

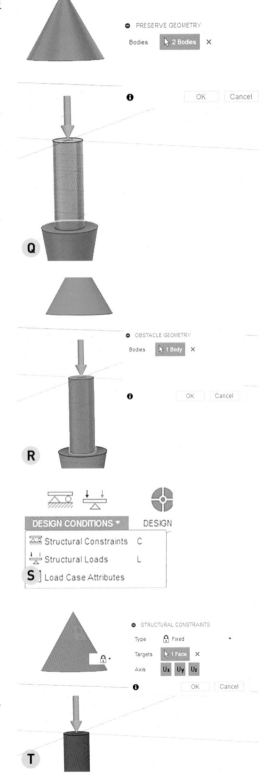

AI to place material. You can also choose a starting shape (this is optional), which is the shape the AI starts from.

- **Preserve Geometry:** Select the body or bodies you want to remain untouched (Figure **Q**). You don't need to hold the Shift key down for multiple selections. I selected the base and the cone. The preserved shapes turn green after you click OK.

- **Obstacle Geometry:** Select the bodies you want the material to go around; that is, to not place any new material in that area. I selected the cylinder (Figure **R**). It will turn red after you click OK.

- **Starting Shape:** This is optional. Click on the body you want to serve as a jump-off point for the AI design. Nothing in this case was selectable.

## DESIGN CONDITIONS

Apply structural constraints and loads to your selected geometry (Figure **S**).

- **Constraints:** These can be *fixed*, *pin*, or *frictionless*. You can apply constraints to multiple objects, and depending on the constraint type, to faces, edges, or vertices. I applied a fixed constraint to the cone (Figure **T**). Note the colors. Yellow is for starting shape (existing

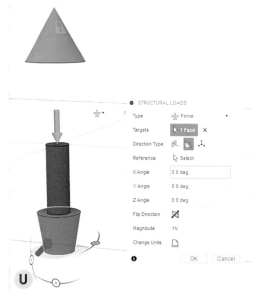

**Loads**

| | Type | Name | Magnitude | S |
|---|---|---|---|---|
| ✓ | Gravity | Gravity | 9.807 m / s^2 | In Cent |
| ✓ | Force | Force1 | 1 N | Attache |

**Constraints**

| | Type | Name | Attributes | S |
|---|---|---|---|---|
| ✓ | Fixed | Fixed1 | Ux; Uy; Uz; | Attache |

**TIP:** Users who need a load-bearing design might want to do a simulation study on the part before giving it to the AI to develop.

model), green is for preserve geometry, and red is for obstacle geometry.

- **Loads:** These simulate pushing, pulling, and twisting forces that your design should withstand. You can apply loads to multiple objects, and depending on the load type, to faces, edges, or vertices. I applied a load to the base and entered 1N (newton; a unit of force) in the dialog box (Figure **U**).
- **Load case attributes:** This is a dialog box that shows what you've applied (Figure **V**).

## DESIGN CRITERIA

Specify data that will help achieve the goals you want for the outcomes (Figure **W**).

- **Objectives:** What do you want optimized? I chose to minimize mass (Figure **X**).
- **Manufacturing:** Choose from Unrestricted, Additive, and Milling manufacturing methods (Figure **Y**). You can select all options and compare outcomes generated for each option.

## MATERIALS

Click on Manage Physical Materials (Figure **Z**) and then choose at least one material to use in the design process. An outcome is generated for each material you choose.

- **Study Materials:** Here you can pick a manufacturing method and a material from the built-in library (Figure **Aa**).
- **Manage Physical Materials:** This opens the Material Browser (Figure **Bb**), where you can manage and edit material libraries, choose favorites, modify properties, and create materials. I dragged the Aluminum icon up to the In This Study window to apply it to the design.

## GENERATE

Pre-check, preview, generate designs, and find the generate status and details here (Figure **Cc**).

- **Pre-check:** This checks the active generative study to ensure the setup meets the requirements to generate outcomes. You might get a dialog box listing warnings and errors to fix before continuing. You may be able to continue with warnings (they're coded yellow), but not with errors (they're coded red). But if everything is fine, you'll get a message saying the study is ready to generate.

☑ Unrestricted

▼ ☑ **Additive**

Orientation  [X+] [Y+] [Z+]

X−  Y−  Z−

Include all six directions  ☐

Overhang Angle  45.0 deg

Minimum Thickness  3.00 mm

▼ ☑ **Milling**  ◄ Milling box

Configuration 1  🔧 3-axis

+  ✕

Tool Direction  X+  Y+  Z+

X−  Y−  [Z−]

Include all six directions  ☐

Minimum Tool Diameter  10.00 mm  ◄ Tool size

**Y**  ol Shoulder Length  40.00 mm

▼ In This Study

Methods  All methods

🔘 Aluminum AlSi10Mg

▼ Library

Library  Fusion 360 Material Library

📁 Ceramic
📁 Electronics
📁 Fabric  **Drag**
📁 Flooring
📁 Gas
📁 Glass
📁 Liquid
📁 Metal

🔘 60Sn40Pb

🔘 63Sn37Pb

🔘 Aluminum

**Bb**

**TIP:** If you try to generate a study and get an error message that the milling tool is too large for the model, change the size in the Minimum Tool Diameter box or just uncheck the Milling box.

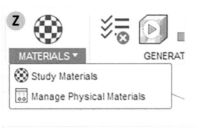

**Z**

MATERIALS ▼  GENERAT

⊗ Study Materials

Manage Physical Materials

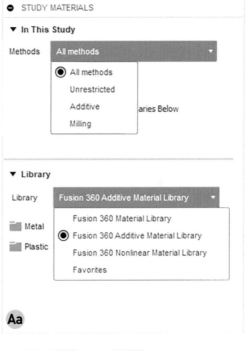

● STUDY MATERIALS

▼ **In This Study**

Methods  All methods ▼

🔘 All methods
  Unrestricted
  Additive  aries Below
  Milling

▼ **Library**

Library  Fusion 360 Additive Material Library ▼

📁 Metal    Fusion 360 Material Library
🔘 Fusion 360 Additive Material Library
📁 Plastic   Fusion 360 Nonlinear Material Library
  Favorites

**Aa**

GENERATE ▼

⊗ Pre-check

▶ Previewer

⬆ Generate

☰ Generate Status

ℹ Generate Details

**Cc**

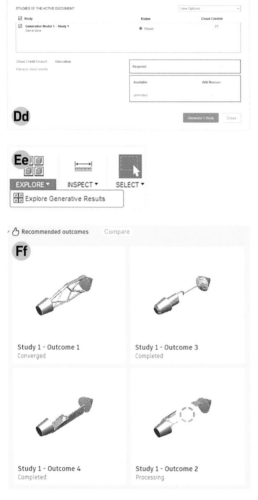

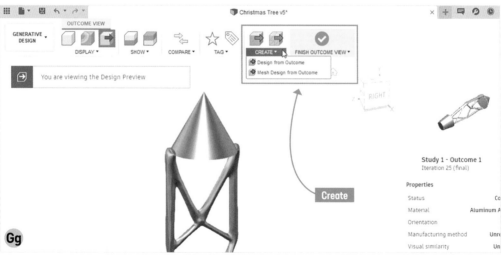

- **Previewer:** This generates an outcome preview after Pre-check verifies the study is ready to go, and before you run the job. You can use this to see how your setup affects outcomes, and adjust accordingly, if needed. Previewer ignores manufacturing and material. Previews can take hours to appear, and Fusion is unusable during that time. It doesn't require cloud credits.
- **Generate:** This designs outcomes that meet your design requirements (Figure **Dd**). It's cloud-based and requires cloud credits. You can select multiple studies. Track the processing status in the Job Status dialog box. The amount of time it takes depends on how many loads you've applied and the overall complexity, but 3–4 hours or longer is normal. You can exit Fusion and the project will still generate.
- **Generate Status :** This displays a list of completed jobs and jobs in progress. You can cancel a job before one iteration of any outcome is generated.
- **Generate Details:** This displays information about outcome generation that you can save to a log file.

### EXPLORE AND CREATE
Click the Explore icon to view your outcomes (Figure **Ee**). You can filter and compare multiple outcomes (Figure **Ff**). Click on a thumbnail to see a larger version (Figure **Gg**). The Create menu lets you choose to output a Design (solid) model and a Mesh model (Figure **Hh**). After closing Fusion, you

can return to the Explore icon to review the outcomes again.

## EDIT THE GENERATIVE OUTPUT

### THE DESIGN MODEL

This consists of a T-spline body and the preserved geometry. A boundary fill is done automatically. However, the model will generally need more editing.

You may be able to edit it in the Form space by right-clicking on a form icon in the Timeline and choosing Edit. But if that doesn't work — an edit option doesn't appear, or the model is ghosted in that workspace — bring it into the Solid workspace. Then click on each Timeline icon from left to right. They'll highlight specific geometry and temporarily put that geometry in the Form workspace to edit (Figure Ii).

### THE MESH MODEL

Bringing this mesh model into the Mesh workspace is a bit counter-intuitive because typically the Timeline must be turned off. Right-click on the Browser root and choose Capture Design History. Then right-click on the mesh and choose Edit (Figure Jj). This will take you into the Mesh workspace (Figure Kk). Tweak your mesh model, then export it for 3D printing and see what your generative design looks like in real life!

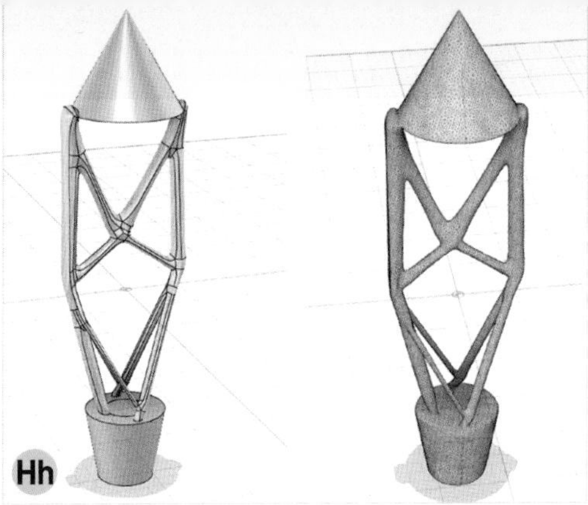
Hh

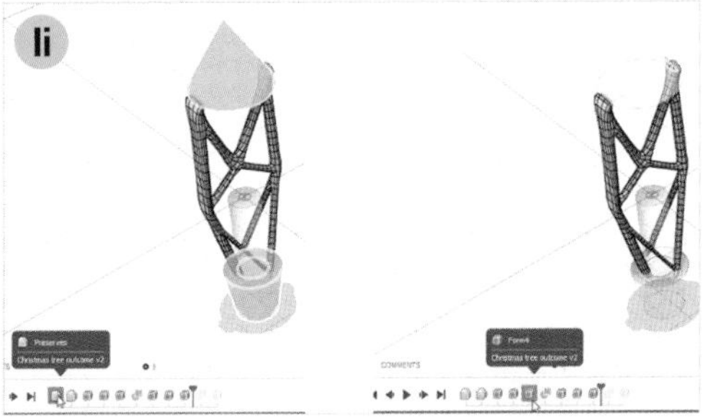
Ii

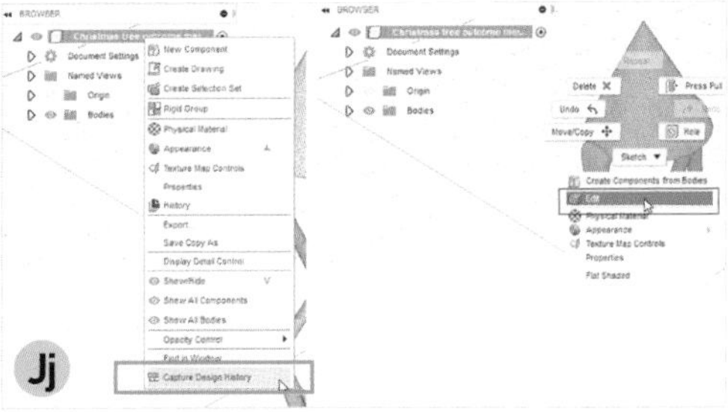
Jj



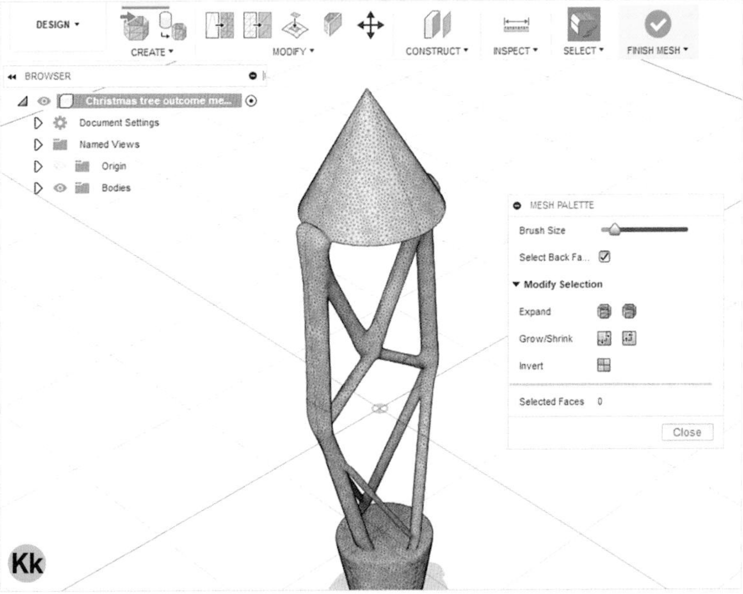

Kk

# ANTENNAS
# IN THE WILD

## Get the skinny on radio antennas with this primer and field guide Written by Tim Deagan, KJ8U

There are easily thousands of different types of antennas serving different purposes. Some are simple lengths of wire, others exotic assemblages of high math and science. Most of the antennas around us are pretty straightforward and have a set of identifiable characteristics. With a few patterns in mind and some rules of thumb, you can usually make a pretty good guess at what most antennas are being used for, or at least the parts of the EM spectrum on which they're operating.

**TIM DEAGAN**
likes to transform things from the digital world into real life in Austin, Texas.

## MAKING ELECTROMAGNETIC WAVES

Antennas are driven from radios by feedlines consisting of two conductors (often bundled in a coax cable with a center conductor and a shield) which carry the two sides of the alternating current signal the radio produces. The simplest antenna, a **dipole**, takes the two conductors from the feedline and spreads them out away from each other. The antenna might appear to be a break in the circuit since the two arms of the antennas are separated instead of connected. What's actually happening is that the two wires are acting like a big air-gap capacitor. Instead of the current being contained within the dielectric gap of a capacitor, the current is flowing in space as an electric field arc between the two arms of the dipole (Figure **A** ).

As the oscillating AC signal moves electric charges back and forth across the antenna arms, the accelerating charges create a magnetic field around the antenna, just like an electromagnet. Without trying to go too deep into Maxwell's equations, rest assured that changing magnetic fields create electric fields, and changing electric fields create magnetic fields. The electrical energy applied to the antenna is thus converted into electromagnetic waves that self-sustain (magnetic and electric fields each relentlessly creating the other) and radiate away from the antenna (Figures **B** and **C** ). Radio!

This only works efficiently if the antenna has the appropriate length for the frequency of signal it's trying to radiate. An antenna that is perfectly matched to a frequency is called **resonant**. When you change the free length of a guitar string by pressing on frets, you change the note (frequency) the string resonates at. When you change the

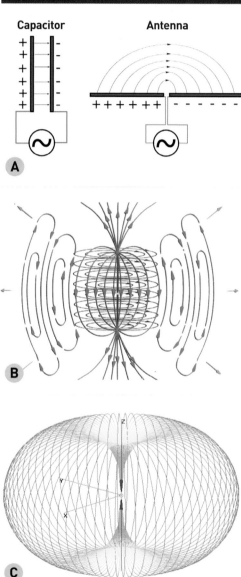

length of an antenna, you change the frequency that it resonates at (though *resonance* has a number of additional technical meanings when used in electronics and radio).

# FREQUENCY AND WAVELENGTH

**Frequency** is the number of times the wave completes a cycle in a given amount of time. The common unit of measure is the Hertz (Hz), or cycles per second. Since waves travel through space at the speed of light, the distance between peaks, called the **wavelength** ($\lambda$), is directly related to the frequency. This is generally measured in meters and derived by dividing the speed of light by the frequency. A signal with a frequency of 14.074MHz has a wavelength of 21.03m.

Ranges of frequencies are usually referred to as **bands**, such as the US 40m amateur radio band (7.000MHz–7.300MHz) or the US AM radio band (525kHz–1705kHz).

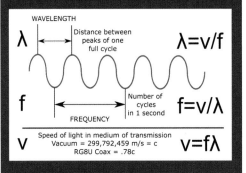

WAVELENGTH

$\lambda$ — Distance between peaks of one full cycle — $\lambda = v/f$

$f$ — Number of cycles in 1 second — $f = v/\lambda$

FREQUENCY

$v$ — Speed of light in medium of transmission
Vacuum = 299,792,459 m/s = c
RG8U Coax = .78c — $v = f\lambda$

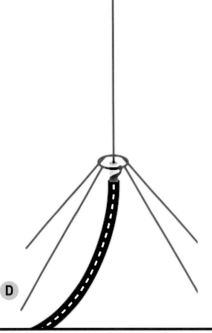

**D**

Tim Deagan, Wikimedia Commons: Psgs123xyz, Chetvorno

## ANTENNA SIZE AND SHAPE

Antenna lengths are generally whole-number multiples or fractions of a target wavelength ($\frac{1}{2}\lambda$, $\frac{1}{4}\lambda$, $\frac{1}{8}\lambda$, $1\lambda$, $2\lambda$, $4\lambda$, etc.), so they often provides clues about the intended frequency of operation. Long wavelengths/low frequencies tend to use long antennas and short wavelengths/high frequencies tend to use short antennas. Engineers have created a wealth of tricks to allow antennas to operate on multiple bands, but there are almost always limits. Shortwave antennas designed for wavelengths of tens of meters aren't very useful for PCS cell wavelengths in the tens of centimeters, and vice versa.

If your initial image of an antenna is the wire sticking out of your car, don't worry! That single wire (a **monopole** instead of a dipole) is doing the same thing as one arm of the dipole (Figure **D** ). As a substitute for the other arm, monopole antennas connect that side of the feedline to a reference source, or **ground plane**, that's usually perpendicular to the monopole (shown below the radiation pattern in Figure **E** ). The ground plane could be your car, a set of wires, the earth, or in the case of cell phones, our bodies.

## SHAPING ELECTROMAGNETIC RADIATION

Monopoles and dipoles send their signals out in a donut-shaped 360° ring around the length of the wire. This is referred to as an *omni-directional* signal. However, electromagnetic (EM) waves interact with matter in different ways. In some cases it's absorbed, in some reflected, and others refracted. Conductive items that respond to EM waves may, in turn, re-radiate them out again. If the secondary radiation interacts with the initial waves, the two signals may reinforce or cancel each other out depending on their phase relationship. This is referred to as *constructive* and *destructive interference* (Figure **F** ).

When carefully managed, this effect can be used to shape and direct the output of an antenna. Many antenna designs have a driven element called an **emitter** (frequently a dipole) and one or more additional elements (**reflectors** and **directors**) that shape the resulting combined field. While no antenna can add power to the initial signal, by using additional elements a

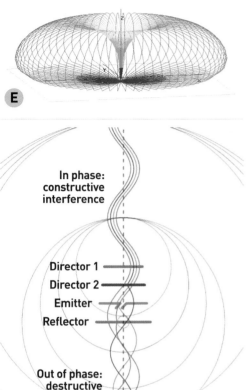

**E**

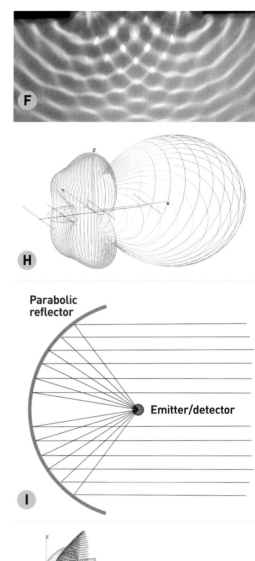

**F**

**In phase: constructive interference**

Director 1
Director 2
Emitter
Reflector

**Out of phase: destructive interference**

**G**

**H**

**Parabolic reflector**

Emitter/detector

**I**

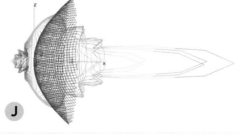

**J**

power gain can be created in one or more specific directions by reducing power in other directions. The result is often called a *uni-directional* signal (Figure **G**). These are described as **beam** or **directional** antennas and are either fixed at a specific target to improve transmission and reception or rotated to change targets (Figure **H**).

EM waves' ability to reflect can be leveraged to great effect when they are aimed at a **parabolic dish**. Satellite dishes and space telescopes both use the same principle of concentrating a tight capture area into a small point for reception and focusing the output of a small point into a directional beam for transmission. The dish is a passive reflector element and the driven element is positioned at the focal point of its parabola (Figures **I** and **J**).

Reflection also allows EM waves to be routed through shaped hollow pipes called **waveguides**. Though there are examples of waveguides for all kinds of waves, these tools are a common

feature used in antennas used for extremely high-frequency waves, commonly called *microwaves* when above 1GHz (though other definitions start microwaves at 300MHz). Waveguides allow a low-loss way to direct EM waves while also serving as a *high-pass filter*, rejecting waves below a certain frequency (Figure **K** on the following page).

K

L

M

x

Wikimedia Commons: Chetvorno, Tim Deagan

Figure A's diagram illustrating the electrical wave arching from one end of a dipole to the other implies an orientation that antennas create for the waves they generate. This effect is called **polarization** and it is an important aspect of antenna design. Antennas have different polarizations if they are mounted vertically or horizontally. A receiving antenna operates best if it has the same orientation as the transmitted signal it's picking up. Other polarization schemes are used as well. **Helical antennas** are an example of a circular polarization design that can receive both horizontal and vertically polarized waves. This is especially useful for receiving signals from satellites (Figure **L**) that are changing orientation (and therefore polarization) as they orbit or experiencing an effect called *Faraday rotation* as their signal traverses the ionosphere (Figure **M**).

Most antennas operate by interacting with the electrical field of an EM wave. But the magnetic field is always there too, and that's the aspect used to transmit and receive with most **loop antennas**. These antennas can be circular, hexagonal, square, or any closed geometry. There may be more than one loop serving the same purpose as the additional elements on a beam antenna. Small loop antennas (Figure **N**) can have especially sharp *nulls* where they can't receive. This makes them especially useful for radio direction finding. By turning the loop until a signal disappears, you can determine the direction from which it is coming (Figure **O**).

## HOW ELECTROMAGNETIC WAVES TRAVEL

EM waves can travel, or **propagate**, in a variety of ways, though different frequencies have different abilities in this regard. These propagation types are: **ground waves**, **direct waves**, and **sky waves** (Figure **P**). Extremely low-frequency (used for communication with submarines around the world) to medium-frequency signals (3Hz–3MHz, such as AM radio) hug the Earth as they travel and are called *ground waves*. High-frequency (HF) waves in the 3MHz–30MHz range (and medium-frequency waves) have the ability to refract off the layers of the ionosphere as *sky waves*, if they're at a shallow enough angle (otherwise they head into space), often multiple times, and reach around the world. Very high-frequency (VHF) and above waves (30MHz and up) rely on *direct*, or *line-of-sight*, transmission. These waves are obstructed by the curvature of the Earth or other obstacles and tend to have limited range compared to lower frequencies. Of course, that depends on where you point them. Aim them at a satellite and even a handheld radio can contact low Earth orbit!

The higher an antenna is mounted, the farther its line-of-sight radius is. Tall antenna towers provide broad coverage for antennas relying on direct propagation. Cellphone towers, commercial radio, amateur radio repeaters, microwave data transmission, emergency services, and other UHF/VHF/microwave antennas all tend to be mounted as high as feasible to increase their range.

Even HF antennas attempting to bounce signals off the ionosphere benefit from height. Some part of the signals they output bounce off the Earth below them. When they are less than half a wavelength above the ground, these reflected waves interfere with the transmitted wave and tend to make most of the signal go upward at an angle so steep that it either passes through the ionosphere or bounces straight back down. This means that an antenna operating on the 40-meter amateur band would ideally be at least 20 meters (65 feet) above the ground to achieve long distance propagation. Figures **Q** and **R** show the comparison of this band being broadcast at 5 meters high and 20 meters high. In some cases, if you want to communicate locally, aiming up and bouncing down is desirable. This is referred to as **near-vertical-incidence skywave (NVIS)** operation.

All these characteristics provide clues when spotting an antenna in the wild. The length of the elements relates to the frequency of operation. The height offers clues to the coverage area. Odd shapes like horns suggest waveguides which are most likely used for high-bandwidth microwave transmissions. Directional antennas are usually pointed at a target such as a receiver at a given compass bearing. If they're pointed at the sky, that target is probably a satellite.

We spend our lives constantly bathed in the radio waves propagating around us. Much of our

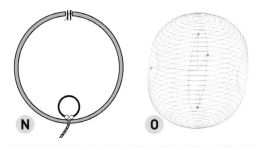

**N**　　　**O**

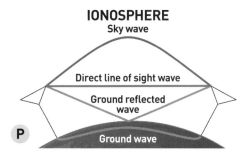

## IONOSPHERE
### Sky wave

Direct line of sight wave

Ground reflected wave

Ground wave

**P**

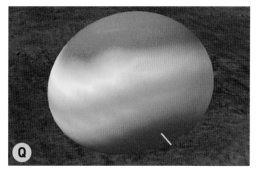

**Q**

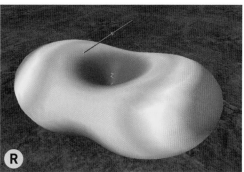

**R**

daily activity relies on radio-enabled cellphones, tablets, GPS, and Wi-Fi. Recognizing the amazing proliferation of antennas can be an exciting way to gain awareness of these tools for manipulation of the invisible forces underlying our modern world. And with that, let's take a look at some of the more common antennas you'll see in the wild ...

# ANTENNA SPOTTING FIELD GUIDE

Antennas are all around us, but what does each kind actually do? Here are a variety of different types, and how they're used. See how many you can find near you!

## DIPOLE

Dipoles consist of two arms, each connected to one side of the radio circuit feedline. Can be constructed from stretched lengths of wire or rigid metal tubing. "Rabbit ear" TV antennas are a common example. Used for nearly every radio band. Size ranges from huge to tiny. Particular favorites of the **amateur radio** aka **ham** community.

A **Horizontal, Vertical, and Diagonal** — Mounted horizontally, a dipole radiates broadside to the length of the wires, with *nulls* (directions in which it can't receive) at either end. Mounted vertically, a dipole radiates in an *omnidirectional* 360° pattern (since the nulls point up and down). When one end is mounted higher than the other, it's called a *sloper* and the radiation pattern changes with the angle of slope.

B **Inverted V and L** — Dipoles are often mounted with the highest point in the center and the two arms angled down, usually at a 120° or 90° angle. *Inverted-vee dipoles* provide a more omnidirectional pattern than a horizontal. Dipoles can also be mounted with one arm horizontal and one vertical, in an *L-dipole* configuration. This is usually due to space constraints.

C **Folded** — When the far ends of the dipole are connected with a parallel wire, the result looks like a flattened oval and is called a *folded dipole*.

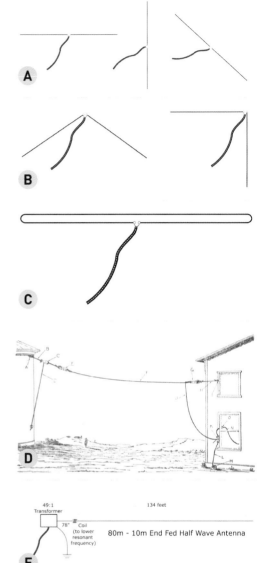

## MONOPOLE

Extremely common anywhere that room for a dipole isn't available. Frequently used in commercial radios to save space. If one of the arms of the dipole is replaced with a connection to a ground plane (wires, car, person, earth), the antenna becomes a monopole.

**D** **Long Wire** — While they require a tuning circuit to work well, long lengths of *random wire* can be used as antennas .

**E** **End-Fed Half Wave** — By using a transformer to connect to a wire that's a half-wavelength long, the antenna will be resonant on the primary wavelength plus its harmonics, making it a *multi-band* antenna.

**F** **Car** — Car radio antennas are usually around 75cm long, approximately one quarter of the center wavelength of the FM radio band (2.78m–3.41m).

**G** **Handheld** — Walkie-talkies, handheld radios, and older cellphones typically have antennas sticking out of them. They may be telescoping whips or wire coiled in a plastic extrusion. Our bodies serve as their ground planes!

## DIRECTIONAL

These allow for transmission/reception toward specific locations, while lowering functionality in other directions. Generally referred to as *beam* or *directional* antennas.

**H** **Yagi** — While not all antennas with a central shaft and parallel extrusions are specifically *Yagi* antennas, the term often gets used generically for this and similar styles of directional antenna (including *Uda* and *log-periodic* types). Yagi-style antennas are loved by hams for the gain they offer. High-end broadcast television antennas are often Yagi antennas since they can be aimed at a stationary broadcasting station.

**I** **Hexbeam** — The hexbeam directional antenna uses a W-shaped driven element in front of a U-shaped reflector element, all arranged as

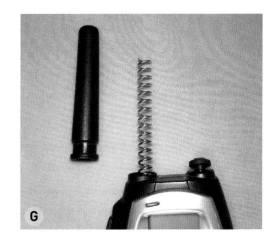

**G**

**H**

**I**

Tim Deagan, Wikimedia Commons: Henry Smith Williams, Zuzu, Shootthedevgru, Merja Partanen, Greg Goebel

a hexagon. Hexbeams are a favorite of ham radio operators for 20m–6m operation. Longer wavelengths make the construction too large.

**J Parabolic (Dish)** — A transmitting/receiving element placed at the focal point of a parabola collects and emits focused radio waves from the dish. Commercial and government installations often use parabolic antennas, as does satellite TV.

**K Waveguide** — These are channels that direct and focus radio waves in the microwave range. Frequently used for high-speed digital signals between buildings or relay towers.

**L Helical** — Antennas produce polarized radio waves. These can be horizontal, vertical, or circular depending on the orientation of the emitting elements. Orbiting satellites change position, and therefore polarization, so a helix does a good job of dealing with those changes. Helical antennas are used whenever it's difficult to get the receiver oriented to the same polarity as the transmitter.

**M Cellphone Tower Sector Antennas** — Cellphone tower antennas are designed to work in a very narrow angle, or *sector*. Sets of these provide 360° coverage without sacrificing signal strength. Sector antennas are the most common configuration for cell service since they provide excellent coverage.

## TINY AND HUGE

Other types of antennas, from PCBs in personal devices to massive pieces of engineering used for commercial, military, and scientific purposes.

**N and O 2.4GHz (Bluetooth, Wi-Fi, etc.) PCB Antennas** — Some devices operate on wavelengths short enough to allow traces on a printed circuit board to serve as antennas. The zig-zag trace shown in Figure N is a Bluetooth 2.4GHz (12.5cm) antenna along with the ground plane of the board. The ESP8266 board includes a built-in Wi-Fi antenna (Figure O), extremely common in IoT and mobile devices. You probably have a few around you right now!

**P** **High-Band (Millimeter Wave) 5G Antennas** —
Very tiny, used by 5G cellular devices operating
near 30GHz (10mm).

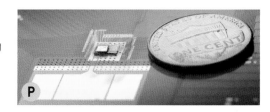

**Q** **ALLISS** — Commercial, government, and
high-power shortwave stations with coverage
that needs to be changed directionally may
use giant, steerable *ALLISS* antennas (from
the French towns of Allouis and Issoudun).

**R** **"Elephant Cage" (AN/FLR-9, AN/FRD-10)** —
During the Cold War, massive ring antenna
systems were set up to monitor and locate
radio transmissions worldwide. The last of
these "elephant cage" antennas can still be
seen in Canada and Alaska.

## LOOP

Unique in that many of them respond to the
magnetic rather than the electrical portion of the
electromagnetic waves. Have very sharp nulls,
making them useful for direction finding. Very
portable; common in receivers and low power
(QRP) operations.

**S** **Circular** — Loop antennas can be circular,
rectangular, or any closed geometry. They
are often easier to deploy than large wire
antennas.

**T** **Quad** — Loop antennas can also include
multiple elements that modify their directional
transmission and reception abilities.

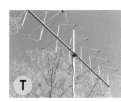

**U** **Radio Direction Finding Loop** — *Radio direction
finding (RDF)* usually uses small receiving
loops that are inefficient for transmitting, but
have extremely sharp nulls to assist in finding
the direction of a radio source. ✪

David Baillot/UC San Diego Jacobs School of Engineering; Wikimedia Commons: Eyreland, Chaddy, Trixt, Daderot; with permission Randy Taylor, WB5QFM

Shop our Toolbox recommendations:
makezine.com/go/toolbox

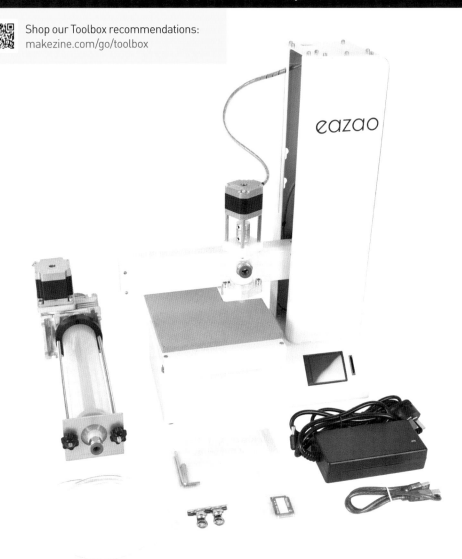

## Eazao Ceramic 3D printer $399 eazao.com

This 3D printer is specifically designed for making ceramic creations, and compared to what we've seen in this space previously, it gives an incredibly affordable option for that medium (most of the machines we've seen that do ceramics have an additional zero on their price tag).

Despite its low price, the Eazao is very capable. The model I got uses a geared-down stepper motor to push clay into an auger type extruder. This means constant pressure, no air systems to deal with, and repeatable and reliable prints. Keep in mind though, you will have to learn all the ins and outs of working with ceramics now, including how to "fire" it, and how to clean up all the material. –Caleb Kraft

 **See our in-depth look at the machine and video of it in action: youtu.be/56Cv3HEnIvQ**

## Lotmaxx Shark V2

**$499** lotmaxx.com | amzn.to/3xdQMoa

The Shark V2 has the ability to use two filaments for multi-color or even multi-material 3D prints. It also has a swappable laser-engraving head. Unfortunately, the laser isn't particularly powerful and its software feels incomplete. As a printer, though, the V2 works great. —*Caleb Kraft*

 **See our in-depth look at the machine and video of it in action: youtu.be/Wbn-RoNjaLs**

## Claw Gardening Gloves

**$4** amzn.to/36kZnK5

Digging through dirt is a devilish treat with these garden gauntlets. They also perfectly complete *any* Halloween costume. The claws are hot-glued on, so don't leave them sitting in the sun. —*Mike Senese*

## AxiDraw MiniKit 2

**$325** shop.evilmadscientist.com

The AxiDraw pen plotter is a simple writing and drawing machine that's available in several versions. Developed by our friends at Evil Mad Scientist with British maker Lindsay Wilson, this mini version is space-saving and portable. The 6"×4" work area is sufficient to write on postcards or envelopes, to sign documents or artworks in series, or to aid people who can't use their hands but still want to send "handwritten" notes (these can be generated from engraving fonts, aka single line fonts, on the PC). You can also use the plotter to transfer digital works of art with pens onto paper, using ordinary vector graphics files.

Assembling this kit requires only scissors and a few screwdrivers. A wide variety of pens can be clamped into the plotter, and you're free to choose any paper — you can mount smaller papers to the included easel board, or just place the plotter on top of larger papers. It can also write on material that's not attached to it at all, such as the tabletop. Plotting is easily controlled by a set of extensions for the free vector graphic software Inkscape. —*Elke Schick*

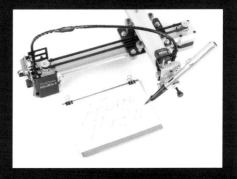

## FormArt 2 Vacuum Former $2,999 myyardtech.com/en

Vacuum forming is one of those capabilities that is very simple in principle, but it has many opportunities to become a pain in practice. Getting your heat, timing, and suction all tuned in for optimal results can really be difficult.

The FormArt automates many of these issues to help you be more productive. They sell material with tiny QR codes; the machine scans these codes and determines the settings that will work best for you. Of course, you still have the capability of manually entering temperatures and timing for whatever material you have on hand.

The system has everything built in, so there's no external pump or multiple plugs for heaters. It really is a one-stop vacuum forming station. FormArt 2 funded on Kickstarter this spring; you may have a waitlist ahead of you as they fill orders. —*Caleb Kraft*

 **See our in-depth look at the machine and video of it in action:** youtu.be/wULPmI_Dcvs

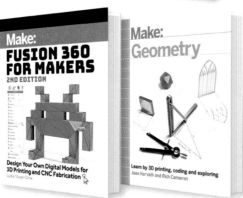

## MyBar Cocktail Machine

**$299** mybar.io

I first saw this project at Maker Faire and thought it looked really fun. I finally got a chance to build one myself and I have to say it delivers on its promise. The kit becomes a handy robot that can mix and dispense various concoctions based on common ingredients, effectively creating a robot bartender. Building the kit is simple enough, primarily involving mounting various pumps and sensors to a custom-cut shell. The real fun happens when you boot up the app (which feels well made) and select a drink. MyBar buzzes to life and spits out the proper amounts of each liquid to get you your tasty treat.

The only downside won't come as a surprise to anyone who has worked in restaurants or bars: You're going to have to clean this thing frequently to keep it from becoming a sticky mess that attracts bugs. —*Caleb Kraft*

## Adafruit Funhouse

**$35** adafruit.com

This house-shaped microcontroller combines temperature, barometric pressure, humidity, and light sensors, giving you the basics of a weather station or room monitor. With its built-in TFT display you can quickly get a read out of any sensor. For a simple project, or when debugging and developing, this display might be all you need. There are also 5 RGB LEDs and a buzzer to alert you to what the board is up to, and pinouts for digital, analog, and $I^2C$ sensors, allowing you to make alarm systems, doorbells, light control systems, and more.

The board is designed for Circuit Python; Adafruit provides a library just for the board allowing for easy programming of its sensors, screen, and three capacitive-touch buttons. It is also possible to flash an Arduino program. Keep in mind that the Funhouse's processor is the new ESP32-S2 that supports Wi-Fi like its predecessor, but not Bluetooth. —*Kelly Egan*

**Find more *Make:* board reviews at makezine.com/comparison/boards.**

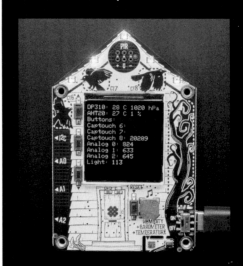

# OVER THE TOP

# UNNECESSARY INVENTIONS

HANDYCHOPS

POWERNAP DESK

MEGAPOD

SWIPE N LIKE

Real-life product designer **Matt Benedetto** spends his off time dreaming of outlandish contraptions and then brings them to life as one-offs, all the way through glossy promotional photo shoots for his site unnecessaryinventions.com. From utensils to apparel to furnishings to electronics, they're all something you might really want, but nothing you'll ever really need. —*Mike Senese*